Free Logic

Selected Essays

Free logic is an important field of philosophical logic that first appeared in the 1950s. Karel Lambert was one of its founders and coined the term itself.

The essays in this collection (written over a period of forty years) explore the philosophical foundations of free logic and its application to areas as diverse as the philosophy of religion and computer science. Among the applications are those to the analysis of existence statements, to definite descriptions, and to partial functions. The volume contains a proof that free logics of any kind are nonextensional and then uses that proof to show that Quine's theory of predication and referential transparency must fail.

The purpose of this collection is to bring an important body of work to the attention of a new generation of professional philosophers, computer scientists, and mathematicians.

Karel Lambert is Research Professor of Logic and Philosophy of Science at the University of California, Irvine.

Free Logic

Selected Essays

KAREL LAMBERT

University of California, Irvine
University of Salzburg

 CAMBRIDGE
UNIVERSITY PRESS

CAMBRIDGE UNIVERSITY PRESS
Cambridge, New York, Melbourne, Madrid, Cape Town, Singapore, São Paulo

Cambridge University Press
The Edinburgh Building, Cambridge CB2 8RU, UK

Published in the United States of America by Cambridge University Press, New York

www.cambridge.org
Information on this title: www.cambridge.org/9780521818162

First published 2003
This digitally printed version 2007

A catalogue record for this publication is available from the British Library

Library of Congress Cataloguing in Publication data

Lambert Karel, 1928–
Free logic: selected essays / Karel Lambert.
p. cm.
Includes bibliographical references and index.
ISBN 0-521-81816-8
1. Free logic.   I. Title.

BC129.L34 2002
160–dc21

2002020177

ISBN 978-0-521-81816-2 hardback
ISBN 978-0-521-03922-2 paperback

*To my dear Carol, whose beauty, sweetness of nature, capacity to laugh,
and unfailing dignity have sustained me for more than half a century.*

Contents

Introduction

In one way or another the nine chapters of this book all have to do with free logic. Most are updated revisions in and adaptations of previously published papers. The exceptions are Chapters 4 and 5, though Chapter 4 contains a revised segment from a recently published paper.

Chapter 1 began as an invited address to the Western Division of the American Philosophical Association Meetings in 1991, and at the request of the organizers of those meetings was subsequently published in slightly revised form in *Philosophical Studies*, 65 (1992), pp. 153–167. It is a critical analysis of Russell's famous theory of definite descriptions of which there are two quite distinct versions. The defining feature of either version is that definite descriptions are not singular terms. That the essence of Russell's theory has to do with logical grammar was stressed in my 'Explaining away singular existence statements', *Dialogue*, 1 (1963), pp. 381–389, and later, independently, by David Kaplan in 'What is Russell's theory of descriptions?' *Physics, History and Logic* (eds. W. Yourgrau and A. Breck), Plenum Press, New York (1970), pp. 277–288.

Chapter 2 is an adaptation of several papers. The main essays are 'Existential import revisited', *Notre Dame Journal of Formal Logic*, 4 (1963), pp. 288–292, 'Notes on E! III: A theory of descriptions', *Philosophical Studies*, 13 (1963), pp. 5–59, 'Notes on E! IV: A reduction in free quantification theory with identity and definite descriptions', *Philosophical Studies*, 15 (1964), pp. 85–88, and 'Free logic and the concept of existence', *Notre Dame Journal of Formal Logic*, 8 (1967), pp. 133–144.

This chapter lays out a motivation for, and the first axiomatic formulation of, a (positive) free logic. It presents the original semantical foundations, and applies that logic to the analysis of singular existence and definite descriptions. Indeed, it contains the first consistent and complete free theory of definite descriptions.

Chapter 3 is essentially an adaptation of two essays. The first is 'On the reduction of two paradoxes and the significance thereof', and appeared in a volume in honor of Gerhard Scheibe entitled *Physik, Philosophie, und die Einheit der Wissenschaftne* (Hrsg. Lorenz Krüger and Brigitte Falkenburg), Spectrum: Heidelberg (1995), pp. 21–33. The second is 'A theory about logical theories of "expressions of the form 'the so and so' where 'the' is in the singular"', and appeared in a memorial issue of *Erkenntnis* (35 [1991], pp. 337–346) in honor of Rudolf Carnap and Hans Reichenbach. This chapter shows how two paradoxes discovered by Russell, one in Meinong's theory of objects and the other in a Frege-inspired formulation of set theory, stem from the same source, the naïve theory of definite descriptions. It uses this information to provide an explanation of the origins of the various traditions in the treatment of definite descriptions exactly parallel to the explanation of various approaches to set theory in the wake of Russell's paradox. Finally, it details some benefits of the free definite description theory approach especially as they concern the logics of definite descriptions in Russell and Frege.

Chapter 4 is new in this book. It is a formal and philosophical examination of the original (informal) theory of definite descriptions of David Hilbert and Paul Bernays, and also of certain neo-Hilbert-Bernays approaches, especially one due to Sören Stenlund. I am very much indebted to Paul Schweizer for his help in the formalization of the *original* Hilbert and Bernays theory. The critique of Stenlund's approach is my own, and Schweizer is blameless. The upshot is that Stenlund's theory, despite his claim to the contrary, is not a free definite description theory, but an interesting development hovering somewhere between the original Hilbert-Bernays treatment (as formalized by Schweizer and me) and free definite description theory.

Chapter 5 presents the foundations of much unpublished work in the 1980s and 1990s done by Peter Woodruff and me on what Bas van Fraassen originally called "the spectrum" of (positive) free definite description theories. It lays out the motivation for one approach to the subject and provides a uniform procedure for proving the completeness of various theories in what I prefer to call – less misleadingly – the hierarchy of positive free definite description theories. It is of a piece with Kripke's semantical analysis of the Lewis hierarchy of modal logics.

Chapter 6 is the updated adaptation of two essays. The first is 'Predication and extensionality', *The Journal of Philosophical Logic*, 3 (1974), pp. 255–264, and the second is 'Fixing Quine's theory of predication', *Dialectica*, 52 (1998), pp. 153–161. This chapter contains the proof that Quine's theory of predication (and hence his theory of referential transparency) is non-extensional, a proof to which Quine himself devoted some attention in his post humous essay, 'Confessions of a confirmed extensionalist'. It also seeks to restore the extensional features of the theory without recourse to the elimination of singular terms.

Chapter 7 is a considerably revised version of the essay entitled 'Nonextensionality'. It was published in a volume in honor of Franz von Kutschera entitled *Das weite Spektrum der analytischen Philosophie* (Hrg. Wolfgang Lenzen), de Gruyter, Berlin (1997). It shows how the logical dependence of two notions of extensionality – the truth-value dependence conception and the *salva veritate* substitution conception – can be restored despite a forceful argument that they are not in a language that contains singular terms without existential import.

Chapter 8 is a much revised version of my Collège de France lectures on the philosophical foundations of free logic, lectures I was invited to give in the spring of 1980. They were published in the journal *Inquiry*, 24 (1981), pp. 147–203 under essentially that title. These lectures, awarded the Medal of the Collège de France, comprehensively examined motivations, discussed confusions, considered applications and laid down now widely adopted conventions.

Chapter 9 is a considerable revision of 'Logical truth and microphysics'. It was published initially in a memorial volume in honor of Henry Leonard entitled *The Logical Way of Doing Things* (ed. Karel Lambert), Yale University Press, New Haven (1969). It was the first application of van Fraassen's technique of supervaluations to a topic in the philosophy of science. His method initially was invented to provide a completeness proof for (positive) free logic that did not depend on model structures utilizing inner and outer domains. Chapter 9 shows that van Fraassen's method can also be used to reconstruct the Reichenbachian interpretation of elementary microphysical statements without appeal to a third truth-value along with a defense of that reconstruction against complaints by Michael Scriven and Wesley Salmon, among others.

There are many to thank in the development and preparation of this book. In particular, I am indebted to Paul Schweizer in the formal development of the original Hilbert-Bernays theory of definite descriptions, and to Peter Woodruff in the essay on the hierarchy of positive free definite descriptions theories. I have enjoyed Woodruff's companionship for nearly forty years. Nothing can detract from his enormously powerful intellect and good nature. Whether the huge amount of work we have done together on definite descriptions ever gets into public print is in the lap of the gods. More generally, I am especially indebted to Bas van Fraassen, Robert K. Meyer, and Brian Skyrms for years of friendship and support, and with each of whom I have been privileged to work on various topics. I am also indebted to longtime personal associations with Edgar and Inge Morscher at the University of Salzburg, Wolfgang and Ulli Spohn at the University of Konstanz, and the late Jules Vuillemin of the Collège de France. I should also like to thank John and Julie Trafford, the former for having discovered the Trafford Eclipse, and the latter for restoring the Sun. Thanks are also due to my good friend Lee Sandler for his many kindnesses. To my children, Kal, Kathryn, and Christopher I am grateful for taking it upon themselves to keep my professional efforts in perspective via loving irreverence. For the preparation of the current book, I thank Terence Moore and his colleagues at Cambridge University Press. They are a firm, decisive, and very helpful group of people.

1

Russell's Version of the Theory of Definite Descriptions

1. INTRODUCTION

It is mildly ironic that the title of this chapter is an unfulfilled (or improper) definite description because Russell really had two versions of the theory of definite descriptions. The two versions differ in primary goals, character and philosophical strength.

The first version of Russell's theory of definite descriptions was developed in his famous essay of 1905, 'On Denoting'.[1] Its primary goal was to ascertain the logical form of natural language statements containing denoting phrases. The class of such statements included statements with definite descriptions, a species of denoting phrase,[2] such as 'The Prime Minister of England in 1904 favored retaliation' and 'The gold mountain is gold'. So the theory of definite descriptions contained in what Russell himself regarded as his finest philosophical essay is a theory about how to *paraphrase* natural language statements containing definite descriptions into an incompletely specified

---

[1] Bertrand Russell, 'On Denoting', *Mind*, New Series: XIV (1905), pp. 479–493; Alfred North Whitehead and Bertrand Russell, *Principia Mathematica*, (Second Printing), Cambridge, At the University Press (1910), Volume 1. 'On Denoting' is reprinted in *Bertrand Russell: Logic and Knowledge* (editor, Robert C. Marsh), George Allen and Unwin, Ltd., London (1956), pp. 41–56. All references here to 'On Denoting' are to the reprinted version in the Marsh collection.

[2] Dismissed by G. F. Stout as rubbish, 'On Denoting' was praised by F. P. Ramsey as a paradigm of philosophical analysis. Russell's own opinion of the quality of his famous essay is reported on page 39 in the Marsh collection cited in the previous note.

formal language about propositional functions. Russell used this version of his theory to disarm arguments such as Meinong's arguments for beingless objects. Such reasoning, he said, is the product of a mistaken view about the logical form of statements containing definite descriptions.

The second and later version is presented in that epic work of 1910, *Principia Mathematica* (hereafter usually *Principia*). Its primary goal, in contrast to the first version, was to provide a foundation for mathematics, indeed, to reduce all of mathematics to logic. In chapter *14 Russell introduces a special symbol, the inverted iota, and uses it to make singular term-like expressions out of quasi-statements. They serve as the formal counterpart of definite descriptions, and the expression 'definite description' is extended to cover the formal counterparts themselves, not an uncommon procedure in logic. Then contextual definitions are offered which are said to "define" definite descriptions in all the possible statements in which they can occur. Definite descriptions are regarded not as a referring kind of expression but as a certain variety of "incomplete symbol". So, in *Principia*, Russell's theory of definite descriptions is a theory about how to treat the logical counterpart of natural language expressions of the form 'the so and so' where 'the' is used in the singular. As such it is a *definitional extension* of a formal language, the first order fragment of which is similar to the predicate logic found in most contemporary textbooks of symbolic logic, minus names. Russell uses definite descriptions in *Principia* for all sorts of purposes; for example, he uses them to define descriptive functions.

The chronological order of the two versions will be reversed and the second version will be discussed first. It is the most complicated of the two versions, is more prone to technical complaint, and mainly because of these same complaints, it is weaker in philosophical strength than the first version of the theory. In fact, the first version is a very natural antidote to many of the problems besetting the second version.

2. RUSSELL'S THEORY IN *PRINCIPIA MATHEMATICA*

In what follows Russell's inverted iota is replaced by a smaller case 'i', the dot notation is replaced by parentheses, '&' replaces '.', his sign for conjunction, and the higher case English letters 'P' and 'Q' replace his Greek symbols 'Φ' and 'Ψ'.

Russell's formal theory is captured in the following pair of *contextual* definitions along with an explanation of their distinctive character.[3]

CD1 [ix(Px)] Qix(Px) = (∃y)((x)(Px ≡ x = y) & Qy) Df
CD2 E!ix(Px) = (∃y)((x)(Px ≡ x = y)) Df[4]

In contrast to the contextual definitions of identity (in chapter *13), and of the conditional (in section A of part I), which, Russell says, define the introduced expressions, **CD1** and **CD2** define neither 'i' nor '**ix(Px)**' even though the signs for identity, the conditional and the sign 'i' are not primitive signs. According to Russell, **CD1** and **CD2** merely "define" any "proposition in which [the phrase '**ix(Px)**'] occurs".[5]

The contextual definitions of definite descriptions in *Principia* are very complex. First, they introduce not one but two symbols along with a symbol pair; 'i', 'E!' and the left and right hand brackets '[' and ']'. Second, they introduce the important notion of the scope of a definite description; this is the function of the brackets around an expression of the form '**ix(Px)**' in the definienda of **CD1** and **CD2**. Third, they show that '**E!**', in contrast to predicates such as '**P**' and '**Q**', only appears next to definite descriptions. Fourth, they reveal that definite descriptions occur in positions in statements often occupied by (logically proper) names or variables, and that '**E!**' occurs in positions in statements often occupied by predicates. For instance, in

a = ix(Px),

'**ix(Px)**' occupies a position often occupied by a name or variable, and in

E!ix(Px),

'**E!**' occupies a position often occupied by predicates, for instance, the predicate '**Q**'. Despite this fact, '**E!**' is not treated as a predicate

---

[3] Russell actually gives a third definition in *Principia* for ordering the occurrence of definite descriptions in statements. That definition is neither important nor essential to the current discussion.

[4] 'CD' abbreviates 'contextual definition'. Neither the expression nor the abbreviation appear in Russell's statement of the two definitions in chapter *14 of *Principia*.

[5] Ibid., p. 175.

(primitive or defined), and definite descriptions are not treated as names or referring expressions (primitive or defined).

In this version of Russell's theory, why it would be a disaster to treat definite descriptions as names, as symbols "directly represent-ing... object[s]", is easily answered. If they were so treated, then

ix(Px & ∼ Px) = ix(Px & ∼ Px)

would be a substitution instance of the valid *Principia* principle

x = x

and hence would be true. But the statement in question is false when evaluated via **CD1** because it is false that

(∃x)(Px & ∼ Px).

On page 72 of *Principia* Russell claims to have *proved* on pages 67 and 68 that definite descriptions are not (logically proper) names, hence that they are "incomplete symbols" and do not stand for "constituents" of "facts". The relevant passages are these:

Suppose we say: 'The round square does not exist.' It seems plain that this is a true proposition, yet we cannot regard it as denying the existence of a certain object called the 'the round square'. For if there were such an object, it would exist: we cannot first assume that there is a certain object, and then proceed to deny that there is such an object. Whenever the grammatical subject of a proposition can be supposed not to exist without rendering the proposition meaningless, it is plane that the grammatical subject is not a proper name, that is, not a name directly representing some object.

... By an extension of the above argument, it can easily be shown that [ix(Px)] is *always* an incomplete symbol. Take, for example, the following proposition: 'Scott is the author of Waverley'. [Here "the author of Waverley" is [ix(x wrote Waverley)].] This proposition expresses an identity; thus if "the author of Waverley" could be taken as a proper name, and supposed to stand for some object *c*, the proposition would be 'Scott is *c*'. But if c is any one except Scott, this proposition is false; while if *c* is Scott, the proposition is "Scott is Scott," which is trivial, and plainly different from "Scott is the author of Waverley." Generalizing, we see that the proposition [a = ix(Px)] is one which may be true or may be false, but is never merely trivial, like [a = a], whereas, if [ix(Px)] were a proper name [a = ix(Px)] would necessarily be either false or the same as the trivial proposition [a = a]. We may express this by saying that [a = ix(Px)] is not a value of the propositional function [a = y], from which it follows that [ix(Px)] is not a value of [y]. But since [y] may be anything, it follows that [**ix(Px)**] is nothing. Hence, since in use it has meaning, it must be an incomplete symbol. . . .

Certain peculiarities of, and complaints about, this version of Russell's arise immediately. Among the more striking are the following. First, consider Russell's treatment of singular existence statements. That treatment seems arbitrary and, in a certain sense, appears to reduce the expressive power of this version of his theory. On the one hand,

E!ix(Px)

and

$$(\exists y)(y = ixP(x))$$

are logically equivalent in the second version, where '**P**' is any one-place predicate. The second of these statements, in fact, is another way of asserting existence. So, on the other hand, one would suppose,

E!a

and

$$(\exists y)(y = a)$$

would represent alternative ways of asserting the existence of the object named by '**a**'. But this is not so because though the latter statement in the immediately preceding pair is logically true, the former, according to Russell, is "meaningless". If, to protect this doctrine, '**E!**' is accorded privileged status as the means of expressing singular existence, the decision seems to be simply an arbitrary syntactical choice with no substantive explanatory power. Moreover, in Spinoza's philosophy, Substance is the one and only one thing that exists. But the natural paraphrase of this descriptive phrase is not well formed in Russell's *Principia* theory because it would juxtapose '**E!**' to a variable as in

$$ix((y)(E!y \equiv y = x)),$$

a juxtaposition which Russell regards as meaningless. So the second version lacks a certain expressive power.[6]

---

[6] Russell's view of the meaninglessness of '**E!x**' is stated on pages 174–175 of *Principia Mathematica*.

Second, Russell's use of definite descriptions in *Principia* is anomalous. Consider, for instance, the definition of a descriptive function in Chapter 30. That definition is expressed as follows:

R'y = ix(xRy) Df.[7]

However, according to Russell, normally a definition like this one is "a declaration that a certain newly-introduced symbol or combination of symbols is to mean the same as a certain other combination of symbols of which the meaning is already known".[8] As such it serves as justification for the replacement of the definiendum by the definiens in formulae of the formal language. In the preceding definition the "other combination of symbols of which the meaning is already known" is the definite description

ix(xRy).

But, as noted earlier, Russell claims that definite descriptions in the second version do not have meaning in isolation, that they themselves are never defined but only the propositions in which they occur. Because definite descriptions have no meaning in and of themselves, they have no meaning that is already known, as is presupposed in the definition above of a descriptive function. So the use of the definite description in the definiens of the above definition is anomalous since evidently it is not treated there as an incomplete symbol.

Russell is aware of the problem as a look at the bottom of page 232 of *Principia* makes clear. His solution is to say that the definition above of a descriptive function is a definition in a very special sense; it is, he says, "more purely symbolic than other definitions". But a dilemma looms. If the definition of a descriptive function is not a definition in Russell's standard sense, then it is hard to see how it can be used to *justify* the replacement of the definiens by the definiendum in the formulae of *Principia*, as Russell clearly intends. Indeed, in the "purely symbolic sense", the definition amounts simply to an unjustified declaration that the definiens and the definiendum are interchangeable. If, on the other hand, the above definition is taken in Russell's usual sense, then the definiens is *not* being treated as an incomplete symbol. This is dramatic evidence of Russell's vacillation in the second version over

---

[7] Ibid., p. 232.
[8] Ibid., p. 11.

the status of definite descriptions. On some occasions, he treats them as logical subjects and, on others as incomplete symbols, and hence *not* as logical subjects.[9]

Third, Russell's "proof" that definite descriptions in the second version are incomplete symbols is highly questionable. The structure of the proof contained in the previously quoted passages is hard to discern. In particular, it is not clear what Russell means by the phrase "By an extension of the above argument... ". Does this phrase suggest an argument by cases in which, first, unfulfilled definite descriptions are shown to be incomplete symbols, and, then, fulfilled definite descriptions are so shown? Or does it suggest merely that there is a basic kind of context, other than the context of non-existence, namely, the context of identity, in which definite descriptions, fulfilled or unfulfilled, can be shown not to be (logically proper) names? Whatever the exact superstructure of Russell's proof, several of its evident premises are open to serious question.

In the first place, the demonstration that the statement 'The round square does not exist' leads to contradiction if its constituent definite description is treated as a name, as a phrase "directly representing some object", rests on the assumption that being an object entails being an existent. Here Russell, apparently, is making tacit appeal to his earlier 1905 "demolition" of Meinong's view that there are nonexistent objects. But, as current discussion has shown, the most Russell's famous argument in 'On Denoting' established is that the principle,

The so and so is (a) so and so,

is, on its most common construal, false. This principle was indeed espoused by Meinong, but it is modifiable or expungible without damaging Meinong's belief in nonexistent objects.[10] In fact, current philosophical logic abounds in provably consistent treatments of nonexistent objects.[11]

---

[9] It should be observed that the definition of a descriptive function on page 132 does not provide a context – namely, an identity context – in which the definite description occurs. Contexts of the form '... = __ Df' are not identity contexts as Russell points out on page 11 of *Principia.*

[10] See Karel Lambert, *Meinong and the Principle of Independence,* Cambridge University Press (1983), pp. 33–34.

[11] See, for example, Terence Parsons, *Nonexistent Objects,* Yale University Press, New Haven (1980).

In the second place, Russell's assertion that a statement of the form

$a = ix(Px)$'

where 'a' is a logically proper name, is "never merely trivial" is itself false; it suffices to let '$ixP(x)$' be '$(x = a)$'. The resulting statement is just as "trivial" as

$a = a.$

In the third place, it is a controversial matter whether the principle of the *substitutivity of identity* holds in contexts of the form

... is trivial,

a principle apparently exploited by Russell when, in effect, he substitutes the expression 'the proposition that Scott is the author of Waverley' for the expression 'the proposition that Scott is Scott' in the statement 'The proposition that Scott is Scott is trivial'. Moreover, even assuming that the the *substitutivity of identity* holds in contexts of the form

... is trivial,

there is the Fregeian position with which to contend. Such contexts are indirect (*ungerade*) from Frege's point of view and, thus, the expressions replacing '...' in '... is trivial' will refer not to their ordinary references, propositions (qua statements) for Russell, but to their senses. But in contexts of the form

... is identical with _

the expressions in question to refer to their ordinary references, and, hence, 'The proposition that Scott is the author of Waverley is trivial' cannot be derived by the *substitutivity of identity* from 'The proposition that Scott is Scott'. Russell's implicit discontent with this solution again apparently relies on his 1905 argument in 'On Denoting' – the perplexing Gray's *Elegy* argument – that Frege's doctrine of sense and reference is incoherent. This latter argument, however, is at worst dubious and at least very controversial, even on the most sympathetic interpretation.[12] It seems appropriate to conclude that Russell's boast

---

[12] Op cit., 'On Denoting', pp. 48–51. See Alonzo Church, 'Carnap's *Introduction to Semantics*', *Philosophical Review*, III (1943), pp. 256–272; John R. Searle, 'Russell's

in *Principia Mathematica* to have *proved* that definite descriptions are incomplete symbols is exaggerated.

Fourth, and perhaps most importantly, the second version violates a condition on formal languages that Russell himself seems to acknowledge. The condition is that a formal language should not be ambiguous with respect to logical form.[13] One can think of logical form as the way a statement is evaluated for truth-value.[14] For example,

This is duplicitous,

in contrast to

Someone is duplicitous,

is a predication because, in standard semantics, it is evaluated by locating the referent of 'this' and ascertaining whether or not it is a member of the set of things associated with the predicate 'is duplicitous'. Because the word 'someone' doesn't even purport to have a referent, the contrast statement can't be a predication. Of course, this way of talking about logical form is at the very most only implicit in *Principia* because its semantics is never formally specified. Nevertheless, there is little, if any, distortion of Russell's view that formal languages should be, and are easily made to be, unambiguous with respect to logical form. So consider

Qa,

and

Qix(Px),

objection to Frege's theory of sense and reference', *Analysis*, XVIII, (1958), pp. 137–143; and Simon Blackburn and Alan Code, 'The power of Russell's criticism of Frege: 'On Denoting', pp. 48–50', *Analysis*, 38 (1978), pp. 65–77.

[13] Op cit., 'On Denoting', pp. 52–53.

[14] The rough and ready account of logical form adopted here is due to David Kaplan. See his 'What is Russell's theory of definite descriptions?' in *Physics, History and Logic* (editors, W. Yourgrau and A. Breck), Plenum: New York (1967), pp. 277–295. Actually Kaplan's account does raise questions (not especially troublesome in the current discussion). For example, positive and negative free logicians both count 'Vulcan is Vulcan' as a predication, but the former evaluates the statement true while the latter evaluates the statement false. On Kaplan's account these adversaries have different conceptions of the logical form called 'predication'.

where '**a**' is a logically proper name – perhaps, the word 'this'. These two statements appear to have the same logical form; their syntactic structure would lead one to think that they would be evaluated in the same way. But they are not. Were names to be added to *Principia,*

Qa

would certainly be a predication, but not

Qix(Px).

Indeed the latter statement gets evaluated via **CD1,** and, as is evident, is much more complicated in that regard than the former statement. So Russell's second version violates the condition that formal languages not be misleading in their syntax with respect to logical form.

A similar situation arises *vis-à- vis* the pair of statements

E!ix(Px)

and

Qix(Px)

because in the Russellian scheme of things '**E!**' cannot go in all places where the predicate '**Q**' can. For example, it cannot occupy the place of '**Q**' in

Qx

or in

Qa,

where '**a**' is a logically proper name. In *Principia,* Russell says that contexts of the form

E!a,

where '**a**' is a logically proper name, and contexts of the form

E!x

are not meaningful, but contexts of the form

Qa

and

Qx

are. Accordingly, Russell has a different way of evaluating contexts of the form

E!ix(Px),

namely, via the definition **CD2**. Were '**E!**' a genuine predicate, **CDI** would apply and yield

$$(\exists x)((x)(Px \equiv x = y) \ \& \ E!y)$$

thus violating Russell's stricture that the quasi-statement

E!y

is meaningless.

Many of the difficulties adduced above are, in large part, the result of allowing the expressions '**E!**' and '**ix(P(x)**' to occur in the formal language. Sans types it is essentially classical first order predicate logic, given the conditions for a statement containing a definite description being true or false reflected in the contextual definitions **CDI** and **CD2**. (I assume, of course, that in an adequate formalization of the language of *Principia Mathematica* the definiendum of any definition will count among the well-formed expressions of the formal language.) Indeed, their non-appearance in the formal language comports much better with Russell's beliefs that (singular) existence is not a predicate, and definite descriptions are not (logically proper) names. The latter feature especially is characteristic of the theory of definite descriptions contained in Russell's great essay, 'On Denoting', the theory here identified as the first version.

3. RUSSELL'S THEORY IN 'ON DENOTING'

The treatment of definite descriptions in the first version is analogous to Russell's treatment of indefinite descriptions English phrases of the form 'a so and so'. He exploited this analogy explicitly in the chapter entitled *Descriptions* in his *Introduction to Mathematical Philosophy*, though not in the earliest statement of the theory in 'On Denoting'.[15] Russell did not regard indefinite descriptions as referring expressions, in his words, as "logically proper names". Natural language statements containing them are paraphrased into only a roughly specified formal language. Thus, the statement

I met a man,

to use an example of Russell's, gets paraphrased as

($\exists$x) (x is human and I met x).[16]

This statement does 'not' have the form of a predication and does not contain the expression 'a man' as a logically isolable unit, a "constituent", in Russell's language. Moreover, the rule of paraphrase suggested in the previous example does not yield a unique understanding of all natural language statements containing indefinite descriptions.[17] Consider, for example, the natural language statement

A man met every philosopher.

This may mean either

(Ex) (x is a human male & (y) (if y is a philosopher, then x met y)),

or it may mean

(y) (if y is a philosopher, then (Ex)(x is a human male & x met y)).

Construed in the first way, the expression 'a man' may be said to have *primary occurrence* in its host natural language statement. Construed in the second way, it may be said to have *secondary occurrence* (to borrow terminology that Russell employs in his discussion of definite descriptions in the first version).

[15] Bertrand Russell, *Introduction to Mathematical Philosophy*, George Allen and Unwin: London (1919). See the chapter entitled 'Descriptions'.

[16] Actually the statement in question gets paraphrased as 'The propositional function "x is human and I met x" is sometimes true'. But, in *Principia*, the preceding paraphrase would express what '($\exists$x) (x is human and I met x)' means. The choice of the formal language into which the original statement is to be paraphrased does no violence to Russell's purposes in 'On Denoting' or to the claims in the current discussion.

[17] This point has been emphasized by David Kaplan in 'What is Russell's theory of descriptions?' (cf. footnote 14, above.)

These features of Russell's treatment of indefinite descriptions, the most pervasive in modern philosophical logic, are aped in his treatment of definite descriptions except for the exact character of the paraphrase. Thus, definite descriptions are not logically proper names, natural language statements containing them are not predications, logically speaking, and are ambiguous, their appropriate paraphrase depending on whether the constituent definite descriptions have primary or secondary occurrence. That definite descriptions are not (logically proper) names is put simply and elegantly by Russell in the following passage from 'On Denoting':

> a denoting phrase is essentially part of a sentence and does not like most single words have any significance on its own account. If I say 'Scott is a man,' that is a statement of the form '**x** is a man,' and it has 'Scott' for its subject. But if I say, 'The author of Waverley was a man,' that is not a statement of the form '**x** was a man' and does not have 'the author of Waverley' for its subject.[18]

The actual character of the paraphrases of natural language statements containing definite descriptions can be divined from the definition **CD1** in the second version. Russell does not explicitly treat contexts of the form 'The so and so exists' in 'On Denoting', but his view that (singular) existence is not a predicate requires a separate rule of paraphrase contexts such as that reflected in **CD2**.

The key feature of the first version of Russell's theory is its position on the logical form of statements in colloquial discourse containing definite descriptions. This, in turn, depends on the grammatical status of definite descriptions. Definite descriptions are not "logical subjects", names "in the strict logical sense" (referring expressions), as has been emphasized earlier. Ultimately, the only names or referring expressions for Russell were the demonstratives 'this' and 'that'. Treating definite descriptions as name-like, he thought, would lead to paradoxes of the sort he believed he found in Meinong's theory, a theory that does treat definite descriptions as playing a referring role.[19] So, in *Principia*, he sought to *prove* that definite descriptions are not names. That argument is easily adaptable, of course, to the first

---

[18] Op. cit., 'On Denoting', p. 51.
[19] See, for instance, *The Autobiography of Bertrand Russell; The Middle Years: 1914–1944*, Bantam Books, New York (1969), p. 309.

version, even though the notion of an incomplete symbol does not occur there, because the notion of a name in the strict logical sense already occurs in 'On Denoting'. But if Russell's proof carries over, so do its difficulties. Moreover, just as Russell, in 'On Denoting', dismissed Frege's treatment of definite descriptions as "plainly artificial", so his own treatment can be dismissed as conflicting with presumably natural inferential behavior.

Consider, for example, the reasoning of those astronomers concerned with the planet causing the perturbations in the orbit of Mercury. When this planet, 'Vulcan' by (grammatically proper) name, was discovered not to exist, presumably those same astronomers inferred the nonexistence of Vulcan by a straightforward application of the Principle of the *substitutivity of identity*, despite Russell's opinion to the contrary that definite descriptions are not names (logical subjects). And similarly for other attributions to Vulcan on the basis of like attributions to the planet causing the perturbations in the orbit of Mercury.

Earlier the difficulties enumerated above befalling the second version were said to depend on the occurrence in the formal language of the symbols '**E!**' and '**i**' given the definitions **CD1** and **CD2**. Strictly speaking, however, it depends on what the formal language is and how the symbols above are introduced. It is possible to retain many of the key features of Russell's second version. For instance, agreement with the policy toward the truth-value of statements containing definite descriptions, and his treatment scope, can be had while still rejecting his view that definite descriptions are not referring expressions, *provided* the underlying logic is revised. Paraphrase of natural language statements into the alternative formal language would then have the advantage that natural language statements containing definite descriptions could be treated as genuine predications while nevertheless sustaining Russell's views about the ambiguity of natural language statements containing definite descriptions and the truth values of those statements. This has been hinted at by many people but has been fully developed only by Rolf Schock and Ronald Scales.[20] Finally, the

---

[20] See Rolf Schock, *Logics without Existence Assumptions*, Almqvist and Wiksells, Stockholm (1968), and Ronald Scales, *Attribution and Existence*; Ph.D, Thesis, University of California, Irvine, University of Michigan Microfilms (1969).

preceding remarks establish the general point that there can be a vast difference of philosophical import between the definitional extension *of* a given theory and paraphrases from colloquial discourse *into* that theory.

2

Existential Import, 'E!' and 'The'

1. INTRODUCTION

The traditional logic of general terms supposed the inference from **A** statements to **I** statements to be valid. **A** statements are those treated in the modern logic of general terms (or predicates)[1] as having the form

(1) $\forall x(P(x) \supset Q(x))$.

I statements are those treated in the modern logic of general terms as having the form

(2) $\exists x(P(x) \,\&\, Q(x))$.

But where **P** is replaced by the general term 'perfect frictionless plane', the false statement that there exists a perfect frictionless plane can be inferred. In general, the inference from **A** statements to **I** statements, treated in the manner of the modern logic of general terms, loses its validity when at least the placeholder **P** is replaced by a general term true of nothing. Restricting **P** and **Q** to general terms true of at least one existent thing would restore soundness. Traditionally, such

---

[1] A predicate is what is left of a sentence after a singular term (name or definite description) is eliminated. So, for example, 'is a man' is what remains after deleting 'Adam' in 'Adam is a man'. Here no distinction between predicates and general terms is made though for many only 'man' would qualify as a general term in 'Adam is a man', 'is a' being a stylistic variant of the copula characteristic the medieval logic of terms.

16

general terms were said to have *existential import*,[2] and the *fallacy of existential import* is to infer (2) from (1) in the absence of any restriction on the substituends of **P** and **Q**.

There are two major disadvantages to this way out of the difficulty. First, it unduly limits the scope of application of logic. For example, general terms like 'member of the null class' cannot replace **P**. So logical evaluation of the statement

(3)   The null class is contained in any class

apparently is not forthcoming. Second, it does not permit discrimination of those inferences for whose validity the existence of the things characterized by **P** and **Q** is relevant and those inferences for whose validity their existence is irrelevant.[3]

The modern logic of general terms purports to resolve the matter in another less restrictive way. It allows *unlimited* substitution into the general term place-holders **P** and **Q**, and replaces the inference pattern

(4) $\forall x(P(x) \supset Q(x))$
    $\therefore \exists x(P(x) \mathrel{\&} Q(x))$

by the inference pattern

(5) $\forall x(P(x) \supset Q(x))$
    $\exists x(P(x))$
    $\therefore \exists x(P(x) \mathrel{\&} Q(x))$

So, it would seem, the scope of application of logic is not restricted *vis-à-vis* its general terms, and it becomes possible to evaluate logically statements such as (3). Moreover, it is now also possible to distinguish between inferences whose validity requires that there exist objects of which its constituent general terms are true from those whose validity doesn't. The validity of arguments from statements of the form in (1) to statements of the form in (2) requires a statement of the form

$\exists x(P(x))$,

[2] Ralph Eaton, *General Logic*, Scribners, New York (1936), p. 225.
[3] These points are made with clarity in Henry Leonard's, 'The Logic of Existence', *Philosophical Studies*, 7 (1956), pp. 49–64. This paper is the origin of free logic, though the development therein was not then so called.

but the validity of the argument from a statement of the form in (1) to a statement of the form

$$\sim\exists x(P(x) \text{ \& } \sim Q(x))$$

does not. The soundness of the modern theory of general terms is not compromised by this maneuver. Most contemporary systems of first order predicate logic sans identity, containing a rule of substitution, incorporate something very much like the above solution to the problem initiated by the inference from an **A** statement to an **I** statement and are demonstrably sound.

Curiously, in contemporary first order predicate logic with identity, there is an exception to the inference pattern in (5). Let **P** be the general term of the form

$$= t,$$

where **t** is a singular term (for instance, a grammatically proper name). Then,

(6) $\exists x(x = t \text{ \& } Q(x))$

follows from

(7) $\forall x(x = t \supset Q(x))$

justifying, of course, the inference from (7) to (6). This in turn suggests that modern first order predicate logic treats any general term of the form

(8) $= t$

as having existential import, and, hence that

(9) $\exists x(x = t)$

will always be true no matter what singular term constant replaces **t**. The curiosity borders on the obtuse in view of such apparent counter examples as

(10)   There exists something the same as Vulcan (the putative planet) and

(11)   There exists something the same as the perfect frictionless plane.

The soundness of the modern theory of general terms is restored by restricting replacement of **t** to singular terms with *existential import,* those singular terms actually referring to existing things. But this policy brings with it undesirable features parallel to those noted above in restricting general term placeholders to substituends true of something. For example, the true statement

(12)   Everything identical with the perfect frictionless plane allows unrestricted movement over it,

and the evaluation of its alleged consequences now falls outside the purview of logic. And the capacity to distinguish between inferences whose validity requires the assumption that **t** has existential import and those whose validity doesn't again collapses. So the restriction method *vis-à-vis* singular terms ought to be the course of last resort, and in fact it is.

The most popular way out is to preserve the inference from (7) to (6) and to accommodate statements involving singular term constants via the medium of definite description theory. This course, however, has at least two serious unsatisfactory features. First, it requires treating most if not all grammatically proper names as abbreviations for definite descriptions, an arguable practice.[4] Second, the favored theories of definite descriptions, usually Russell's but sometimes Frege's (*à la* the scientific language), are unproven and/or are unnatural.[5] Here, another way of resolving the problem of existential import posed by singular terms in first order predicate logic with identity is explained. This is a matter of importance if for no other reason than the fact that from a logical point of view identity is the crucial, and, indeed, might be the only (two placed) general term (or predicate), occurring in the language.

2. ANOTHER EXPLANATION

The offending party is not difficult to ascertain in the asymmetrical attitude toward existential import in the modern theory of general

---

[4] See Henry Leonard, *Principles of Reasoning,* Dover, New York (1967), pp. 338–339.

[5] On Russell's theory, see Chapter 1; and on Frege's theory, see Bertrand Russell, 'On Denoting', *Mind,* XIV (1905), pp. 479–493 and reprinted in Bertrand Russell, *Logic and Knowledge* (ed. Robert Marsh), George Allen & Unwin, Ltd, London (1956), p. 47. The reference here is to the reprinted essay in Marsh's collection.

terms. It is that old unreliable, the principle of *Particularization*. Notice that

(13) $\forall x(x = t \supset Q(x)) \supset \exists x(x = t \ \& \ Q(x))$

is deducible from the valid formula of predicate logic,

(14) $\exists x(P(x)) \supset (\forall x(P(x) \supset Q(x)) \supset \exists x(P(x) \ \& \ Q(x)))$,

and detaching with the help of

(15) $\exists x(x = t)$,

a formula deducible from the identity principle

(16) $t = t$,

substituting $= t$, in *Particularization*, that is, in

(17) $P(t) \supset \exists x(P(x))$.

Yet if **t** is replaced by a singular term without existential import in (16) and (17), say, 'Vulcan', and **P** is replaced in (17) by '= Vulcan', one obtains the false

(18) $\exists x(x = \text{Vulcan})$

from the true

(19) Vulcan $=$ Vulcan,

a version of the fallacy of existential import.[6] Moreover, if a predicate of singular existence, say, **E!**, is available in the underlying predicate logic of first order, one need not take the detour through identity theory to show the non-validity of (17). For though it is true that

$\sim$**E!**(Vulcan),

it is certainly false that

$\exists x(\sim$**E!**$(x))$

under the usual interpretation of the quantifiers. These two cases suggest that *Particularization* suffers from a deep disorder. It requires that

---

[6] Op. cit., *General Logic*, p. 225, n. 2.

singular terms have existential import, a dissatisfying story familiar from the previous section. Moreover, one might expect the dual of *Particularization*,

(20) $\forall x(P(x)) \supset P(t)$

the principle of *Specification*, to be subject to a similar malady. The expectation is justified. Let **P** be $\exists y(\ldots = y)$ (or **E!**), and **t** be 'Vulcan'. Then the antecedent of (20) is true and the consequent is false.

It is tempting to think, therefore, that the asymmetrical attitude of the modern theory of general terms toward terms in general is the result of the questionable presumption, already reflected in the laws of *Particularization* and *Specification*, that singular terms have existential import.

Following suggestions in much recent work in the foundations of predicate logic, let us amend first order predicate logic with identity as follows (assuming any adequate system of classical two valued statement logic with '$\sim$' and '$\supset$' as primitive). *Specification* is replaced by

    A1 $\forall y(\forall x(P(x)) \supset P(y))$.[7]

The traditional principle of *universal quantifier distribution*,

    A2 $\forall x(P(x) \supset Q(x)) \supset (\forall x(P(x)) \supset \forall x(Q(x)))$,

is retained. The third principle is a version of the *reflexivity of identity*.

    A3 $t = t$, where **s** and **t** are names,

and the fourth basic principle is a version of the *indiscernability of identicals*,

    A4 $s = t \supset (P(s) \supset P(t))$, where **s** and **t** are names.

The rules of inference are the usual rules for *substitution into predicate placeholders* **P, Q, R,** ... and *singular term (name) place-holders*, **s, t, u,** ... *Modus Ponens* (*Detachment*) and Belnap's rule that

    BR   From any axiom **A**, infer $\forall x A^*$, where **A*** differs from **A** by containing the variable **x** where **A** contains the name **t**.

---

[7] Ermanno Bencivenga has speculated on the deeper philosophical significance of **A1** in 'Why free logic?'. This essay is in his book *Looser Ends*, University of Minnesota Press, Minneapolis (1989).

'Proof', 'Theorem' and 'Consistency' are defined as usual.

All of the theorems of the system sketched hold in all domains including the empty one. To obtain a system whose theorems hold in all domains excluding the empty one, the principle,

A5 $\exists x(P(x) \supset P(x))$,

or its like, must be added. Otherwise theorems like

(21) $\forall x(P(x)) \supset \exists x(P(x))$

are not forthcoming.[8]

These claims, as well as those immediately following, are justified in a semantics whose models are ordered triples of the form $\langle \mathbf{D_O}, \mathbf{D_I}, f \rangle$ where $\mathbf{D_O}, \mathbf{D_I}$ are disjoint sets of objects (possibly empty) (but whose union is nonempty), and **f** is a function – *the interpretation function* – defined as follows:

  (i)   **f(t)** is a member of $\mathbf{D_O} \cup \mathbf{D_I}$, where **t** is a name;
  (ii) **f(P)**, where **P** is an n-adic general term (predicate), is a set of n-tuples of members of $\mathbf{D_O} \cup \mathbf{D_I}$;

and

  (iii) every member of $\mathbf{D_O} \cup \mathbf{D_I}$ is assigned a name.

The definition of *truth in a model* is entirely classical except in the clause for the universal quantifier. That clause reads:

UQ $\forall x(P(x))$ is true in a model just in case **P(t)** is true for all names **t** such that **f(t)** is a member of $\mathbf{D_I}$.

'Validity', 'logical truth' and 'satisfiability' are defined as usual. Intuitively, $\mathbf{D_I}$ is the set of existent objects, and $\mathbf{D_O}$ is the set of nonexistent objects.[9]

---

[8] See Theodore Hailperin, 'A theory of restricted quantification', *Journal of Symbolic Logic*, 22 (1957), p. 31.

[9] The model structure sketched above is essentially the one I presented in lectures at the University of Alberta in the early 1960s. Nuel Belnap, independently, developed an inner domain-outer domain model structure, as this variety of semantics has come to be called, at the University of Pittsburgh during the same period. But the first published accounts of inner domain-outer domain model structures appear in Hugues Leblanc and Richmond Thomason, 'Completeness theorems for some presupposition-free logics', *Fundamenta Mathematicae*, 62 (1968) pp. 125–164, and in Robert Meyer and

The two systems sketched above are easily shown to be semantically sound and semantically complete.[10] The only substantial wrinkle in the key lemma of a Henkin-style completeness proof – that any consistent set of formulas is simultaneously satisfiable – is the requirement that any formula of the form $\exists x(P(x))$ (contextually defined as usual with the help of $\sim$ and $\forall$) be instantiated to a statement of the form $P(t)$ where t has existential import, that is, where $f(t)$ is a member of the set of existent objects D_I.

Particularization does not hold in either of the above systems. Let $f(t)$ not be a member of D_I, but let $f(P) = D_I$, where P is a one-place predicate. Then $\sim P(t)$ is true, but $\exists x(\sim P(x))$ is false. What does hold, and is derivable in either treatment, is a *restricted* version of *Particularization*, that is

RP $\exists x(x = t) \supset (P(t) \supset \exists x(P(x)))$.

RP follows from **A4**, and a deductive consequence of **A2**, namely,

(22) $\forall x(P(x) \supset Q(x)) \supset (\exists x(P(x)) \supset \exists x(Q(x)))$,

via the definition of the existential quantifier. The antecedent of **RP** can be replaced by **E!**(t) given the convention that

(23)   **E!**(t) is short for $\exists x(x = t)$.

Because *Particularization* does not hold, and hence is not a theorem in the systems above, $\exists x(x = t)$ neither holds nor is a theorem. This appears to restore the symmetry of the modern theory of general terms toward the existential import of terms. Moreover, the resulting logic, which was called a *free logic* in 1960, does not require that names

---

Karel Lambert, 'Universally free logic and standard quantification theory', *Journal of Symbolic Logic*, 33 (1968), pp. 8–26. Leblanc, along with Theodore Hailperin, is responsible for one of the first sorties into what Leonard called the logic of existence, and thus deserves recognition as one of the founders of free logic. (See Hugues Leblanc and Theodore Hailperin, 'Nondesignating singular terms', *Philosophical Review*, 68 (1959), pp. 239–243.)

[10] Essentially the first system sketched above, with the addition of an axiom for the commutability of universal quantifiers, (minus the identity theory) is formalized and proved semantically complete via a semantics based on inner domain-outer domain model structures in Hugues Leblanc and Robert Meyer, 'On prefacing $(\forall X)A \supset A(Y/X)$ with $(\forall Y)$: a free quantification theory without identity', *Zeitschrift für mathematische Logik und Grundlagen der Mathematik*, 12 (1970), pp. 153–168. The extension of their proof to classical identity theory is straightforward.

be treated as definite descriptions.[11] Whether they are or are not so treated is a policy not forced on one by purely logical considerations. Finally, it should be clear that the current development of the logic of general terms does not limit the scope of application of logic. It does not exclude from its purview statements such as (12), and it allows one to discriminate between inferences where the existential import of a singular term is crucial (hence, where an existence assumption is required), as in

(24)   **P(t)**
       E!(t)
    ∴ **∃x(P(x))**,

from those in which it is not, as in

(25)   **P(s)**
       s = t
    ∴ **P(t).**

3. ON THE DEFINITION OF E!(t)

In (23) a definition of **E!(t)** was proposed, a suggestion first made by Hintikka.[12] This definition is unacceptable in the classical logic of

[11] Karel Lambert, 'The definition of E(xistence)! in free logic', *Abstracts:The International Conference for Logic, Methodology, and Philosophy of Science*, Stanford University Press, Stanford (1960). Strictly speaking, the system here sketched is an example of a *positive free logic*. In this version of free logic, some atomic statements containing singular terms without existential import are true, for example, (19) above. When this system appeared in 1963, no other kinds of free logic were yet on the market. But soon there appeared another version of free logic, now known as a species of *negative free logic*, by Rolf Schock, 'Contributions to syntax, semantics, and the philosophy of science', *Notre Dame Journal of Formal Logic*, 5 (1964), pp. 241–289. A negative free logic is one in which all atomic statements containing at least one singular term without existential import fails. A consequence of this logic is that *Particularization* holds though *Specification* fails when the premise in the first case, and the conclusion in the latter case, is atomic. Moreover, the identity theory is non-classical because A3 fails. Both kinds of free logic have been exploited in the development of theories of partial functions adequate to programming languages. For an example of the first kind, see Raymond Gumb and Karel Lambert, 'Definitions in nonstrict positive free logic', *Modern Logic*, 7 (1997), pp. 25–55 (For a corrected version see gumb@cs.uml.edu); for an example of the second kind, see Solomon Feferman, 'Definedness', *Erkenntnis*, 43 (1995), pp. 295–320.

[12] Jaakko Hintikka, 'On the logic of existence and necessity I: Existence', *The Monist*, 50 (1966), pp. 55–76.

general terms. In that theory all instances of $\exists x(x = t)$ are logically true. So **E!(t)** would be logically true no matter what name replaced **t**. What makes (23) a legitimate candidate for the definition of **E!(t)** in free logics is the failure of $\exists x(x = t)$ when **t** lacks existential import, for example, where **t** is 'Vulcan'. The question arises whether the definition is justified in free logic, and, if so, whether any other predicate[13] other than one of the form $\exists x(x = ...)$ would suffice. The latter question is answered in the negative later in this book[14]; the former question is answered in the affirmative below.

The sense of the quantifiers in free logic is the same as that in the classical logic of predicates. If a primitive existence symbol, **E!**, is available, this can be made explicit by adding to the set **A1–A4**.

   (A6) $\forall x(E!(x) \supset P(x)) \supset \forall x(P(x)).$

If one wishes to exclude the case where $\mathbf{D_I}$ is empty, **A5** must be added Because (23) is no longer available, neither

   $\mathbf{RP_E}$ $E!(t) \supset (P(t) \supset \exists x(P(x)))$

nor its dual

   $\mathbf{RS_E}$ $E!(t) \supset (\forall x(P(x)) \supset P(t))$

is deducible. Therefore, $\mathbf{RS_E}$ is adopted here as **A7**. It and **A6** may be regarded as Carnapian meaning postulates for **E!** (the singular existence predicate).

On the semantic side, $\mathbf{f(E!)} = \mathbf{D_I}$, and a statement of the form **E!(t)** is true when $\mathbf{f(t)}$ is a member of $\mathbf{D_I}$ (intuitively, the set of existents), and otherwise is false. The expanded system is simply a conservative extension of the original system and hence is easily shown to be sound and complete. A proof of

   $E!(t) \equiv \exists x(x = t)$

is readily available, justifying the convention adopted in **(23)**.

---

[13] The question is vocabulary relative. If, for example, a free logic is supplemented by a term forming operator generating complex predicates out of open sentences, then there is an alternative way of defining 'E!t', as Arthur Prior and Frederick Fitch may have been the first to notice

[14] See Chapter 8.

The proof sketch is as follows:

Case 1: $E!(t) \supset \exists x(x = t)$
  - (a) $t = t \supset (E!(t) \supset \exists x(x = t))$ – **A7 and** *Tautologies*;
  - (b) $E!(t) \supset \exists x(x = t)$ – (a) **and A3**.

Case 2: $\exists x(x = t) \supset E!(t)$
  - (a) $\sim E!(t) \supset (s = t \supset \sim E!(s)$ – **A4**
  - (b) $\sim E!(t) \supset \forall x(x = t \supset \sim E!(x))$ – (a), **BR and** *Quantfier Confinement*;
  - (c) $\sim E!(t) \supset (\exists x(x = t) \supset \exists x(\sim E!(x)))$ – (b) **and A2**;
  - (d) $\exists x(x = t) \supset E!(t)$ – (c), **A6 and** *Tautologies*.

The theorem follows from **Case (1)** and **Case (2)** by the definition of the biconditional. Thus the close tie between the general term $E(xists)!$, identity and the classical understanding of the quantifiers is made explicit in free logic. Moreover, as Hintikka first suggested, the theorem above appears to be the obvious object language counterpart of Quine's dictum that to be is to be the value of a bound variable, at least where singular terms are concerned.[15]

4. DEFINITE DESCRIPTIONS

Turning from names (simple singular terms) to definite descriptions (complex singular terms), the question arises what the reasoning based on free logic involving definite descriptions should look like. Certainly it will depart in important ways from Russell's theory of definite descriptions. Contra Russell, definite descriptions, fulfilled or unfulfilled, are treated as genuine singular terms because any statement of the form 'The so and so is such and such' is regarded a genuine predication. As Quine has urged, treating 'The so and so is such and such' and 'The so and so is not such and such' as nonpredications, when 'the so and so' does not have existential import, is a vestige of the old confusion between the meaning of an expression and its reference.[16]

Though no subscriber to the present liberalized view of predication, it is also the case that in Frege's treatment based on classical first order

---

[15] See the reference in footnote 12.

[16] Willard Van Orman Quine, *From a Logical Point of View*, Harper, New York (1963), p. 163.

predicate logic all definite descriptions are singular terms. But there is a significant semantical difference between free definite description theory and Frege's approach when in his scientific mood. A definition is needed to make the point clear.

A *definite description* is *fulfilled* if its *basis* – the general term (predicate) to which the singular term forming operator 'the (object) **x** such that' – is prefixed, is true of exactly one existent object; otherwise it is *unfulfilled*. (Symbolically, 'the (object **x**) such that' is rendered here as **ix**)

Now in Frege's theory, even unfulfilled definite descriptions have referents among the existents ($\mathbf{D_I}$), but not in definite description theory based on free logic.

Moreover, there is a departure from Russell's misleading transformation rules (theorems).[17] For example, from Russell's definition of $\mathbf{Q(ix(P(x)))}$ in *Principia Mathematica* *14.01, the theorem

(26) $\mathbf{Q(ix(P(x)))} \supset \mathbf{\exists x(P(x)}$

is derivable. But replacing $\mathbf{Q}$ in (26) by

$= \mathbf{ix(P(x))},$

one can infer

(27) $\mathbf{\exists x(P(x))}$

with the help of **A3**. However, (27) becomes false when **P** is replaced by the general term 'is a nonspherical spheroid'. Another consequence of *14.01 is

(28) $\mathbf{Q(ix(P(x)))} \supset \mathbf{\exists x(\forall y(P(x) \supset x = y))}.$

From (28), it is equally easy to obtain

(29) $\mathbf{\exists x(\forall y(P(x) \supset x = y))},$

which is false when **P** is replaced by 'is a writer'. So, in (positive) free logic, (27) and (29) will force the abandonment of the definition *14.01 in *Principia*. It also supports the intuition of some ordinary language philosophers (for example, Strawson), that simple (or atomic)

---

[17] See Chapter 1.

statements containing definite descriptions neither imply existence nor uniqueness of the definitely described object.

So in reasoning based on free logic one can (at least) entertain certain "natural" conjectures that, in *Principia*, hold only under restriction. These natural conjectures are

(30) $P(ix(P(x)))$

and

(31) $ix(P(x)) = ix(P(x))$.

(30) and (31) hold in *Principia* if and only if the condition that

(32) $E!(ix(P(x)))$,

obtains.

In Hintikka's proposal,[18] (30) is a theorem; it follows from his proposed axiom

(33) $t = ix(P(x)) \equiv \forall x((P(x) \supset x = y) \& P(y))$.

Letting **s, t, u** ... of the free logic sketched above in Section 2 be extended to allow them to be replaced either by names or definite descriptions, the proof sketch of (30) is as follows.

  (a) $ix(P(x)) = ix(P(x)) \supset \forall x((P(x) \supset x = ix(P(x))) \& P(ix((x)))$
        From (33), by substituting $ix(P(x))$ for **t**, and *Tautologies*;
  (b) $P(ix(P(x)))$
        From (a), **A3**, *Modus Ponens* and *Tautologies*.

Russell's definition of statements of the form

(34) $E!(ix(P(x)))$

in *Principia* (*14.02) is quite acceptable as a principle in free logic. Now Leonard has noted[19] that (30) yields, by substitution

(35) $\exists y(y = ix(\exists y(y = x)))$.

[18] Jaakko Hintikka, 'Towards a theory of definite descriptions' *Analysis*, 19 (1959), pp. 79–85.

[19] Henry Leonard, 'The logic of existence', *Philosophical Studies*, 4 (1956), pp. 49–64.

Hence, by the definition in (23),

(36)   **E!t $\equiv \exists y(y = t)$,**

and thus

(37)   **E!(ix(E!(x))).**

Russell's *14.02 applied to (37) yields

(38)   **$\exists x(\forall y(E!x \equiv x = y))$,**

a theorem declaring that there is exactly one existent. Except for Spinozists and some Absolute Idealists, (38) is very hard on the intuition. In his own free theory of definite descriptions, Leonard avoids (38) by putting a modal restriction on (30):[20]

(39)   **$\sim\forall x(\sim\diamond\sim P(x) \supset E!(x)) \supset P(ix(P(x))$.**

Nevertheless other more serious problems associated with (30) arise in free definite description theory, whether that theory be modal, as preferred by Leonard, or non-modal as preferred by Hintikka and me.

5. THE DEFINITE DESCRIPTION PARADOX

To the man on the bus nothing could be more obvious than the truth – indeed, the *logical truth* – of (30). But if **P** in (30) is replaced by the general term, 'is round and non-round', then (30) has the ruinous result that

(40)   The object that is round and non-round is round, and, also, is non-round.

One of the virtues of Russell's theory is that the paradoxical (30) is not derivable; the most that can be derived in his theory is

(41)   **$\sim$E!**(the object that is round and non-round).

(40) amounts to a proof of the inconsistency of Hintikka's non-modal theory of definite descriptions. In virtue of (39), Leonard's modal theory avoids this result. The question arises whether a sound theory

[20] Ibid., p. 62.

of definite descriptions, absent modal complications, can be given. The answer is yes.

Hintikka's proposal (33), reminiscent of Peano, would not fail on the assumption that **t** exists, that is, when (33) is replaced by

(42) **E!t ⊃ (t = ix(P(x)) ≡ (∀x(P(x) ⊃ x = t) & P(t))).**

But, in the current free logic, (42) is a deductive consequence of

(43) **∀y(y = ix(P(x)) ≡ (∀x(P(x) ⊃ x = y) & P(y))),**

via the restricted form of *Specification* and the definition in (23).

The potential paradox associated with proposal (33) may be avoided in the same way the noxious consequences of the classical principle of *Specification* were avoided, namely, by adopting the universal closure of (33) as the basic axiom of a free theory of definite descriptions. That principle is simply (43), re-christened here as the axiom

  A8 **∀y(y = ix(P(x)) ≡ (∀x(P(x) ⊃ x = y) & P(y))).**[21]

A useful consequence of **A8**, in the system of free definite description theory being developed here, is the principle

(44) **E!ix(P(x)) ≡ ∃x(∀y(P(y) ≡ y =x)),**

a close syntactic counterpart of Russell's definition *14.02 in *Principia*.

Another important principle evidently not deducible from **A8** is the principle of *Cancellation*:

  A9 **t = ix(x = t).**

The reason is clear; **A8** tells one nothing about the behavior of definite descriptions without existential import except that

(45) **~E!ix(P(x) & ~P(x)).**

A9, in contrast, does. For no matter what singular term replaces **t** in **A9**, whether it be a name or a definite description, **A9** says that the definite description **ix(x = t)** can be cancelled in favor of **t**, even if **t** lacks existential import.

---

[21] Known now as Lambert's Law in the community of free logicians.

6. ADDITIONAL SEMANTICAL CONSIDERATIONS

A question naturally arises about the reference conditions for definite descriptions in the current free definite description theory. How is

f(ix(P(x))

to be treated, and how are statements containing definite descriptions to be evaluated for truth value – especially because the theory has been advertised as being non-Fregeian?

The answer to the first question is clear; add to the description of the models for the free logic sketched earlier the following condition:

(DD)   If $\forall y(P(y) \equiv y = t)$ is true where **t** is a name, then $f(ix(P(x))) = f(t)$, but if $\sim\exists x(\forall y(P(y) \equiv y = x))$ is true, then $f(ix(P(x))) = \mathbf{d} \in \mathbf{D_O}$.

This free interpretation rule for definite descriptions differs from the usual Fregeian rule only with respect to what is assigned to an unfulfilled definite description. Whereas the latter always assigns an arbitrarily chosen *existent* (a member of $\mathbf{D_I}$) to unfulfilled definite descriptions in the most common form of the theory, the former assigns an arbitrarily chosen *nonexistent* to such definite descriptions. One important result is that statements of the form

E!(ix(P(x))

can justifiably be treated as in Russell's natural proposal in *14.02, a policy reflected in theorem (44) above.

The conditions under which statements containing definite descriptions are true require no further expansion or restrictions, when **s**, **t**, and **u** ... are extended to definite descriptions in the free logic developed earlier. It is easy to see that the **A8** and **A9** hold in the current semantical development, but it is also the case – as this machinery shows – that the preceding free definite description theory is incomplete. For

(46) $(\sim\exists y(y = ix(x = t))\ \&\ \sim\exists y(ix(P(x)) = y)) \supset ix(P(x)) = ix(x = t)$

is logically true in the current semantics, but is not derivable in the current proof theory. The easiest way to restore completeness, given the guidance of *Cancellation*, is to append to the underlying free logic with identity the extensionality principle,

A10 $(\sim\exists y(y = t) \ \& \sim\exists y(s = y)) \supset s = t.$

A10 says that if **t** and **s**, whether names on definite descriptions, lack existential import, a statement of the form **s** = **t**' will always be true – in the spirit, but not the letter, of the scientifically inspired Frege.

The free theory of definite descriptions presented above is subject to remarkable proof theoretical simplification. Consider a free logic with **A1**, **A2** and **A4**, and if one wishes, **A5**. Append to that basis

FD2 $ix(P(x)) = t \equiv \forall y(t = y \equiv (P(y) \ \& \ \forall x(P(x) \supset t = x))).$

This new formulation is deductively equivalent to the earlier development, and hence **A4**, **A8**, **A9** and **A10** can be reduced to the status of derived principles. This is interesting from a technical viewpoint, but probably not otherwise. So the earlier formulation of free definite description theory will be the official proposal here. It is in the earlier theory that **A8** is stressed. By semantic descent, **A8** is the intuitive answer to the question "To what (existent) does the 'the so and so' refer?" It says, "To whatever (existent) is such that it and it only is so and so". This very natural principle does not hold in the classical logic of terms, as the reader may quickly verify.

3

The Reduction of Two Paradoxes and the Significance Thereof

1. THE CONJECTURE

David Kaplan once suggested to me that the pair of self-contradictory statements:

(1) The round square both is and isn't a round square,

and

(2) The class of all classes not members of themselves both is and isn't a member of itself

"ought to have the same father". But apparently they don't despite their family resemblance. Russell deduced (1) from a principle he presumed correctly to be a key ingredient of Meinong's theory of objects. That principle says:

MP The object that is so and so is (a) so and so.

On the other hand, Russell deduced (2) from a seemingly unrelated but no less fundamental principle in Frege's version of set theory, the principle of set abstraction.[1] That principle, a version of the principle

---

[1] See William Hatcher, *Foundations of Mathematics*, Saunders, Philadelphia (1968), p. 90. Strictly speaking, the principle in question is not Frege's famous fifth law about classes, but rather is easily derivable from it given a proper treatment of sentences of the form **a** ∈ **{x: A}**. This was first pointed out to me by Terence Parsons, who also asserted that Russell's "deduction" of the famous paradox in Frege's theory needed some help from Frege himself.

of comprehension, (in effect) says:

FP Everything is such that it is a member of the class of
     so and sos if and only if it is (a) so and so.

The lack of common ancestry between **MP** and **FP**, and hence between their respective consequences (1) and (2), enabled Russell to
treat the theory of objects and the theory of sets (or classes) very differently. He thought (1) "demolished" the theory of objects, but he
didn't think (2) destroyed the theory of classes. Russell's attitude was
wrong, because Kaplan's suspicion of the common kinship of (1) and
(2) is justified, and the proof of this fact is the next order of business.
A significant by-product of this result is that the various approaches to
the logic of definite descriptions can be seen as reactions to the failure
of a natural principle about definitely described objects.

2. THE PROOF

Consider the question, "To what does an expression such as

(3)   the Vice-President of the U.S. in 1990

refer?". A natural if somewhat stilted response is, "It refers to whatever
is such that it and it only is Vice-President of the U.S. in 1990". By what
a Quineian would call *semantic descent,* the question and answer give
way, respectively, to the question, "Who is the Vice-President of the
U.S. in 1990?", and the answer, "It is whoever is such that that person
and only that person is Vice-President of the U.S. in 1990."

In symbols, the last answer may be expressed as

(4) $\forall x(x = \imath y(P(y)) \equiv \forall y(P(y) \equiv y = x))$,

where the predicate **P** abbreviates 'Vice-President of the U.S. in 1990'.
Generalizing (4) one obtains

MD $\forall x(x = \imath y(A) \equiv \forall y(A \equiv y = x))$,

where **A** ranges over quasi-statements, that is, sentences, atomic or
complex, containing no free variable but **y** if any variable at all. (The
statements are the closed sentences of the language.)

Suppose now that the generalization, **MD**, of the natural answer symbolized in (4) is added to a first order predicate logic with identity of the sort favored by Quine. This is a logic composed of truth-functions, quantifiers, general terms (or predicates), and identity, but no constant singular terms. **MD** introduces into this otherwise singular term-less language those having the form of definite descriptions – or so I shall assume.

An immediate consequence of this theory – *the naïve theory of definite descriptions* – **NTDD**, for short – is the formal counterpart of the principle **MP**, that is,

(5) **A(iy(A)).**

(5) follows from **MD** by tautologies and the classical logical laws of/ *Specification* and *Reflexivity of Identity*. But, as Russell first noted, an instance of (5) is the disastrous counterpart of (1), that is,

(6) **P(ix(P(x) & ~P(x))) & ~P(ix(P(x) & ~P(x))),**

where **P** is 'is round', and it is assumed that '**t** is square' implies '**t** is not round'. (**t** is a singular term or variable.)

Another immediate consequence of the naïve theory of definite descriptions, provided that the class name **{x: A}** is defined as in

D1 **{x: A}** = df **ix**(∀y(y ∈ **x** ≡ A)),[2]

is the formal counterpart of **FP**, namely,

(7) ∀y(y ∈ **{x: A}** ≡ A).

(7) also follows by tautologies, *Specification* and *Reflexivity of Identity*. It yields the troubling formal counterpart of (2), namely,

(8) **{x: x ∉ x)** ∈ **{x: x ∉ x}** ≡ **{x: x ∉ x}** ∉ **{x: x ∉ x}**,

as Russell also was essentially the first to notice. The upshot is that the generalization of the natural answer in (4), in the context of classical first order predicate logic with identity, yields both **MP** and **FP**, and thus justifies Kaplan's suspicion of their common paternity. At the

---

[2] This definition, as a matter of fact, is favored in Quine's **NF**. It, or at least its syntactical counterpart, where '=**df**' is replaced by '=', is also contained in *20.55 of *Principia Mathematica*.

same time it undercuts Russell's differing attitudes toward Meinongian object theory and a version of Fregeian (or naïve) set theory, the key principles of which are derivable in the common **NTDD**. The derivability of (6) and (8) in **NTDD** also shows that it is a ruined theory. But often ruin leads to insight, and this is one such case.

3. A NEW EXPLANATION OF LOGICAL THEORIES OF DEFINITE DESCRIPTIONS

It will be useful to begin with an analogy from set theory.[3] Different approaches in that theory have often been viewed as reactions to the failure of the naive the principle of set abstraction (comprehension), that is, to the failure of the principle in (7), or some classical equivalent. Zermelo inspired one approach, *amending naïve abstraction*. For instance, Quine, in the spirit of Zermelo if not the letter, replaces the principle in (7) in **ML** by

(9) $\forall y(y \in \{x: A\} \equiv (y \in V \& A))$

(Zermelo did not recognize the universal set **V**.) A second strategy, *restricting the formation rules of the language*, was Russell's own preference. While maintaining (7) he refashioned the logical grammar so that expressions such as $x \in x$ and $x \notin x$ no longer counted as well-formed. A third strategy, *restricting the substitution rules*, is embodied in Quine's system of **N**(ew) **F**(oundations). Though well-formed, expressions of the form $x \in x$ and $x \notin x$ are not permissible substitution instances of **A** in (7) because they lack the property of being "stratified". A fourth suggestion, *rejecting the underlying logic*, is one favored by Intuitionists and Free set theorists, but for different reasons. The former reject the underlying classical logic of statements – in particular the law of *excluded middle*. The latter retain the classical logic of statements and reject the underlying quantifier logic, and specifically the unrestricted principle of *Specification*; specification from the statement of the form $\forall x(P(x))$ is permissible to a purported object **o** in free class theory provided that **o** exists. All of these approaches avoid the unsettling result in (2) (or its formal counterpart (8)).

[3] For the most part, I use the words 'class' and 'set' interchangeably here.

Just as various approaches to the treatment of sets can be regarded as different reactions to the natural if paradoxical principle **FP** (or (7)), so the various approaches to the treatment of definitely described objects can be construed as reactions to the natural if paradoxical principle **MD**. This means that a novel explanation of the various approaches to the logical treatment of definite descriptions is within grasp.

There are four major approaches to the logical treatment of definite descriptions. With one exception, the name of the approach is associated with the name of the inventor or inventors; the exception is the most recent approach, free definite description theory.

The first of these policies, the Frege theory, regards all expressions of the form **iyA** as genuine singular terms and assigns referents among the existents, sometimes arbitrarily, to all of them, including unfulfilled definite descriptions. In particular, it assigns referents even to expressions such as 'the round square' and 'the inept physician'. His theory can be construed as a species of the amendment approach because, in effect, he amends **MD**. That principle is replaced by

$$(10) \quad \forall x(x = iyA \equiv (\forall y(A \equiv y = x)) \; V \; (\sim\exists z(\forall y(A \equiv y = z)) \; \& \; x = *))$$

where $*$ is an arbitrarily chosen object, in one version or $\{y:A\}$ in the other version. Neither **MP** nor **FP** is deducible in this theory except under very restrictive conditions in which all taint of inconsistency is removed. Moreover, Frege's famous elimination principle for definite descriptions in all contexts is straightforwardly derivable from (10).

Another tradition in the logical treatment of definite descriptions originates with Hilbert and Bernays. In effect, they assent to the natural principle of definite descriptions reflected in **MD**, but avoid **MP** and **FP** by restricting the logical grammar of the language. Thus instances of the form **iyA** are counted well-formed only if it is provable that there is exactly one thing such that **A**. (Carnap gave this view a semantical twist in his book *Meaning and Necessity* by substituting the word 'true' for the word 'provable' in this condition on well-formedness.)[4] Thus, the consequences (6) and (8) not only are not derivable in their treatment, they are not even well-formed.

---

[4] Rudolf Carnap, *Meaning and Necessity*, University of Chicago Press, Chicago (1947), pp. 33–34.

Russell invented the third, a version of the most widely adopted treatment. In *Principia* and in the *Introduction To Mathematical Philosophy* expressions of the form **iyA** are regarded as well-formed (even when **A** is non-unique), but they are not regarded as genuine singular terms, as phrases "directly representing objects". So they are not legitimate substitution instances of the free variables in the theorems of the formal system and/or do not fall within the scope of application of certain classical rules of inference. In particular, this restriction on substitution does not permit the inference from **MD** to an instance in which the free variable **x** is replaced throughout by a definite description via *Specification*. As a result, the troublesome **MP** and **FP** are no longer deducible. Indeed, the typographical counterpart of **MD** is a theorem in Russell's system as developed in *Principia*. Russell provides elimination principles for all contexts containing definite descriptions in the early definitions of chapter *14 in *Principia* and it is these which yield a theorem typographically similar to **MD**.

The last, and most recent, treatment of definite descriptions – free definite description theory – treats all definite descriptions as genuine singular terms (like Frege) but does not assign an existent as referent to unfulfilled definite descriptions (unlike Frege). This approach retains the natural principle **MD**, and neither tinkers with the grammar of the language nor restricts its substitution rules. Indeed, **MD** expresses the fundamental property of any free theory of definite descriptions that disallows truth-value gaps (and even of some that do).[5] Rather it rejects the underlying classical predicate logic in favor of a free predicate logic. Hence the appropriateness of the label 'free definite description theory'. It originated with the author in the early 1960s, and other versions by Rolf Schock, Dana Scott, Richard Grandy, Bas van Fraassen and Richmond Thomason, among others, soon followed.

As mentioned earlier (Chapter 2), the principle of *Specification* is rejected and is replaced by

(11) $\forall xA \supset (E!t \supset A(t/x))$,

where **t** is a constant singular term, a grammatically proper name of some kind, a function name or a definite description. (11) permits

---

[5] See Bas Van Fraassen and Karel Lambert, 'On free description theory', *Zeitschrift für mathematische Logik und Grundlagen der Mathematik*, 13 (1967), pp. 225–240.

inference from a statement of universal form to statements containing definite descriptions only on the condition that the described objects exist. So the perfidious statements in (1) and (2) – more precisely, their formal counterparts in (6) and (8) – are derivable only under the unacceptable conditions, respectively, that

(12) $\exists x(x = iy(P(y) \,\&\, \sim P(y)))$

and

(13) $\exists x(x = \{y{:}\, y \notin y\})$

Thus, a uniform way of understanding the different approaches to the logical treatment of definite descriptions is now available.

4. SOME BENEFITS OF FREEDOM

A unique feature of the logical treatment of definite descriptions in free definite description theory is that it permits a comparison of the Fregeian and Russellian logics of definite descriptions without a detour through their respective philosophies of language. If the goal of logic is to determine what statements are logically true, and/or which arguments are valid, then genuine disagreements over these matters require an unequivocal, common idiom. Consider, for instance, the argument from the premise

The round square is the round square

to the conclusion

The round square is both round and square.

This argument gets paraphrased similarly in Frege's and Russell's theories as, in effect,

(14) $ix(R(x) \,\&\, \sim R(x)) = ix(R(x) \,\&\, \sim R(x))$
  $\therefore R(ix(R(x) \,\&\, \sim R(x))) \,\&\, \sim R(ix(R(x) \,\&\, \sim R(x))).$

In Frege's theory (14) is invalid, but not in Russell's theory. Because Russell does not treat definite descriptions as genuine singular terms but Frege does, the disagreement is not genuine. Despite the appearances, the pattern expressed in (14) represents different logical forms in the two logics of definite descriptions. In free definite description theory it becomes possible to examine the logically important disagreements between the Frege and Russell treatments of definite descriptions because such expressions are always treated as singular terms.

In *Attribution and Existence*,[6] Ronald Scales introduced into a negative free logic, a logic in which all simple statements containing singular terms without existential import are counted false, a complex predicate-forming operator λ. When combined with a variable and prefixed to a quasi-statement, it yields a complex predicate, for example,

(15) $\lambda x(x \text{ rotates})$,

which reads 'object (such) that (it) rotates'. The purpose of such an apparatus is to help make scope distinctions. To this end, he adopted the following restricted version of the principle of *predicate abstraction*:

(16) $\lambda x(A), t \equiv (E!t \ \& \ A(t/x))$.

When his negative free logic is supplemented with the axiom schema

(17) $\lambda x(B), iy(A) \equiv \exists z(z = iy(A) \ \& \ B(z/x))$,

and the principle **MD**, basic to virtually all free definite description theories, stylistic variants of Russell's two famous elimination principles for definite descriptions in chapter *14 of *Principia Mathematica* are straightforwardly derivable. In fact, there is exact agreement between the statements containing definite descriptions Russell counts true or false and those Scales counts true or false. For instance, the premise of (14) turns out false in both theories, and thus the argument in (14) is valid in both theories. Moreover, the scope distinctions characteristic of Russell's theory are exactly captured in Scales' free logical development.

On the other hand, it is also easy to reproduce essentially the Fregeian theory in free logic when the principle **MD** is taken as a fundamental principle in the free logic counterpart of Frege's theory. This theory is also well known and goes by the label **FD2** in the literature. To obtain the theory **FD2** it is only necessary to supplement a

---

[6] Ronald Scales, *Attribution and Existence*, University of Michigan Microfilms, Ann Arbor (1969).

positive free logic, a logic in which some simple statements containing singular terms without existential import turn out true, with the basic principle **MD** and the additional schema

(18) $(\sim\!E!(s) \,\&\, \sim\!E!(t)) \supset s = t.$

As mentioned above, the famous Fregeian elimination principle for definite descriptions in all contexts is straightforwardly derivable in this theory.[7] Moreover, just as in Fregeian definite description theory the premise of (14) is not only true, but logically true, so it is in the free definite description theory **FD2**. Hence, in **FD2**, as in Frege's theory, the argument in (14) is invalid because the conclusion is undeniably false.

A further benefit of this analysis is that it brings to light the discovery that an alleged strength of the Russell approach over the Frege approach is simply an artifact of the underlying classical logic, and an important difference vis-à-vis extensionality in the two theories.

Frege's theory is often criticized for its inability to offer a natural treatment of statements of the form

(19)   **E!(ix(A)).**

But when developed in free logic, the Frege policy is not any less natural on this score than is Russell's. For it follows from the basic principle **MD** that

(20) $E!(ix(A)) \equiv \exists y(Ax(Ax \equiv x = y)).$

Because this is one of the major principles in Russell's theory, it follows that the treatment of contexts such as in (20) in the freely developed counterpart of Fregeian definite description theories is no less natural than is the treatment in Russell's theory. This fact is hidden by the development of the Frege's theories in classical predicate logic with identity. Moreover, when the Frege and Russell treatments are developed in free predicate logic with identity, the former theory is extensional in the sense that the object such that it is **P** is identical with the object such that it is **Q** if **P** and **Q** are coextensive, but the latter theory is not. It is sufficient to note that in freely developed Russellian theory

---

[7] Op. cit. 'On Free description theory', *Zeitschrift für mathematische Logik und Grundlagen der Mathematik* (1967).

if the predicates **P** and **Q** are true of no existing objects – or, if one prefers property talk, are possessed by nothing – hence coextensive – it is false that the object such that it is **P** is the same as the object such that it is **Q**. This difference between Russell and Frege is hidden by Russell's decision to disallow definite descriptions in the category of singular term in the logical grammar, a policy forced on him in part by his admiration of classical predicate logic.

Finally, the development of both a Russell-like logic of definite descriptions and a Frege-like logic of same in free logic shows that whether contexts containing negation over a predicate should be regarded as logically equivalent to contexts containing negation over an entire statement does not depend in any way on whether the constituent definite descriptions have existential import. Their non-equivalence, a characteristic of Russell's theory, is reproduced in Scales' free logical counterpart of that theory, and their equivalence, a feature of Frege's theory, is reproduced in **FD2**, the free logical counterpart of Frege's theory. Since definite descriptions qua singular terms can fail to have existential import in either Scales' approach or in **FD2**, this shows that collapse of scope distinctions with respect to negation does not depend upon definite descriptions successfully referring to existents, as indeed they are made to do, in Frege's original theory of definite descriptions. When extended to grammatically proper names, this observation shows that Arthur Prior's suggestion that Russellian names can be identified by the collapse of scope distinctions with respect to negation is incorrect – at least for all languages.[8] For in theories like **FD2**,

(21)   Vulcan is a non-revolver (around the Sun)

and

(22)   It is not the case that Vulcan revolves (around the Sun), derivable, respectively, from

(23)   The object such that it is Vulcan is a non-revolver

[8] See Karel Lambert, 'Russellian Names; Notes on a theory of Arthur Prior' in *Logic and Reality: Essays on the Legacy of Arthur Prior* (ed., Jack C. Copeland), Clarendon Press, Oxford (1996), pp. 411–417.

and

(24)   It is not the case that the object such that it is Vulcan revolves,

in virtue of the logical truth

(25) $t = ix(x = t)$,

are logically equivalent even though the grammatically proper name 'Vulcan' does not have existential import.

4

The Hilbert-Bernays Theory of Definite Descriptions

1. INTRODUCTION

The theory of definite descriptions developed by David Hilbert and Paul Bernays has original and revised versions. The original, and most distinctive version (hereafter **H-BTDD**), received its most explicit statement in the first edition of their treatise on the foundations of mathematics. The account that follows relies primarily on this source.[1] This version of the theory is briefly discussed by Rudolf Carnap, and more fully but informally by G. T. Kneebone and Stephen Kleene (among others).[2] The later theory was Fregeian in spirit, and thus is not distinctive.[3] Moreover, newer versions of the theory (hereafter **Neo-HBTDD** theories), though more in the spirit of the original theory, converge on but cannot be identified with that species called free definite description theory.[4]

---

[1] David Hilbert and Paul Bernays, *Die Grundlagen der Mathematik* I, (first edition), Springer-Verlag, Berlin, 1968. This is a reprint of the same title by Springer-Verlag, Berlin (1934).

[2] Rudolf Carnap, *Meaning And Necessity*, University of Chicago Press, Chicago (1947); G. T. Kneebone, *Mathematical Logic And The Foundations Of Mathematics*, Van Nostrand, New York (1963); and Stephen Kleene, *Mathematical Logic*, Wiley, New York (1967).

[3] David Hilbert and Paul Bernays, *Die Grundlagen der Mathematik* II, (second edition), Springer, Berlin (1939). Reprinted by Springer in 1970.

[4] See, for instance, Sören Stenlund, *The Logic of Description and Existence*, Filosofiska Studier, Uppsala (1973), and Abraham Robinson, 'Constrained denotation' in *Selected Papers*, vol 2, (eds. Jerome Keisler, et. al.), Yale University Press, New Haven (1979), pp. 493–503. (Robinson's original paper was written in 1974.)

Given its essentially mathematical goal **H-BTDD** might be thought to be of limited interest outside logic where the canons of reasoning in any discipline are of concern. But caution in this regard is dictated by the fact that there are modifications of **H-BTDD** where the goal, in part at least, is to provide a treatment of definite descriptions more in keeping with the needs of general philosophy.[5] Moreover, inspired by many of Russell's remarks about (logically proper) names, it is hard to resist to thinking of **H-BTDD** as a theory of (logically proper) definite descriptions.

Hilbert and Bernays note that it is often convenient to introduce into a piece of mathematical reasoning about a specific mathematical object – for instance, a number, a function or a set – an expression referring to that object by means of some uniquely identifying phrase. To this end they propose a notation in which such expressions may be generated out of quasi-statements, for example,

ix(x is between 1 and 3).

They associate such expressions with definite descriptions of colloquial discourse and read the formal counterpart above approximately as

the object **x** such that (**x** is between 1 and 3).

Hence the appropriateness of the label 'definite description' for the formal counterpart.

Hilbert and Bernays do not give a formalized language in which such expressions occur. So the problem arises as how best to represent their view in formal terms. Such a task is constrained by their demand that definite descriptions can be used to make significant statements only if the uniqueness of the *basis* of a given definite description – for instance,

x is between 1 and 3

in

ix(x is between 1 and 3)

---

[5] Ibid., especially Stendlund's monograph.

– is *provable*. This has the effect of placing a restriction on the formation rules of the formal language; expressions of the form

ixA

do not count as (singular) terms in a given language $\mathbf{L_n}$ unless both

∃xA

and

$$\forall x(\forall y((A(x/b) \,\&\, A(y/c)) \supset x = y))$$

are provable in a prior language $\mathbf{L_{n-k}}$.

More informally, the distinctive feature of the original theory is this; for a definite description to count as logically grammatical, the basis of the definite description – 'so and so' in 'the so and so' – must be *provably* unique. This requirement is intended to secure the classical demand that all singular terms, including definite descriptions, have referents.

Two features of the current formalization of **H-BTDD** need to be emphasized. First, the apparent circularity inherent in **H-BTDD** – that the set of logically grammatical expressions of the language depends on what is provable (because what is provable, in turn, depends on the set of logically grammatical expressions) – can be circumvented. The language of **H-BTDD** can be defined in stages, beginning with a definite description free base. Definite descriptions of a given language level can then be introduced on the basis of what is provable in the language of the previous level. All finite depths of embedding of definite descriptions are obtained by defining **H-BTDD** as the union of the results of the procedure just outlined over the finite ordinals. Second, the distinctive feature of the original **H-BTDD** has the consequence that though the ensuing formation rules provide an inductive definition of the language of **H-BTDD**, the set of terms is not decidable because the underlying first order logic is undecidable.

2. SYNTAX OF H-BTDD

1. The **Vocabulary** of **H-BTDD** is as in **NTDD** of Chapter 3.
2. The definite description free base language $\mathbf{L_0}$ formed out of

the vocabulary of **H-BTDD** is the same as classical first order predicate logic.

3. The **formation rules** of **H-BTDD** are these. The *singular terms* of L_{n+1}, formed on the basis of a language L_n, are defined as follows:

(a) If **t** is a singular term of L_n, then **t** is a (singular) term of L_{n+1};

(b) If **A** is a statement of L_n, and if $\exists x(\forall y(A \equiv y = x))$ is provable, then **ixA** is a singular term of L_{n+1}. The *statements* of L_{n+1}, formed on the basis of L_n, are defined as follows:

(c) If A has the form of **s** = **t**, provided **s, t** are singular terms of L_{n+1}, then **A** is a statement;

(d) If **A** has the form of $P^n(t_1,...,t_n)$, provided $t_1,...,t_n$ are singular terms of L_{n+1}, **A** is a statement;

(e) If **A, B** are statements of L_{n+1}, so are $\sim$**A** and (**A & B**);

(f) If **A(b/x)** is a statement **of** L_{n+1}, for all **b**, so is $\forall$**xA**.

4. The usual definitions of **V, ⊃, ≡**, and **∃** obtain, and the definition of *statements* is extended to expressions containing these defined signs.

5. The language of **H-BTDD** can now be now defined as $\cup L_n$, for **n** $\in \omega$.

It is clause **3. (b)** above that differentiates **H-BTDD** from **NTDD** of Chapter 3. The set of expressions of the form **ixA** that count as singular terms in the former is only a subset of those that count as singular terms in the latter. In **H-BTDD**, the pestiferous

Ix(Px & ~Px)

of earlier chapters is not even logically grammatical because, in virtue of **3. (b)**, it does not qualify as a singular term (assuming the consistency of **H-BTDD**).

6. The **transformation rules** of **H-BTDD** are as follows. A statement **A** is *provable* in L_n if it is a consequence of the axioms of L_n by *Detachment*, and an *axiom* of L_n is any tautology or instance in L_n

of the following schemata:

MA1 $\forall x(A \supset B) \supset (\forall xA \supset \forall xB)$

MA2 $A \supset \forall xA$

MA3 $\forall xA \supset A(t/x)$, if **t** is a singular term

MA4 $t = t$, if **t** is a singular term

MA5 $s = t \supset (A \supset A(s//t))$

MA6 $\forall xA(x/a)$, provided **A** is an axiom

MA7 $A(ixA(x/t)/t)$, provided **A** is a statement of $\mathbf{L_{n-1}}$ and $ixA(x/t)$ is a (singular) term of $\mathbf{L_n}$.

If $\mathbf{n = 0}$, then there is no **MA7**, and in the remaining axiom schemata **s** is **a**, and **t** is **b**.

3. SEMANTICS OF H-BTDD

1. A **model** for **H-BTDD** is a pair $\mathbf{<D,I>}$. **D** is a non-empty set, and $\mathbf{I = \cup I_n}$, where $\mathbf{I_i (i \leq n)}$ is an *interpretation* function such that

    (1a) $\mathbf{I_0(a) \in D}$ for any (singular) term **a** of $\mathbf{L_0}$;

    (2a) $\mathbf{I_0(P^n)}$ is a set of n-tuples of elements of **D** for any n-adic predicate of $\mathbf{L_0}$;

    (3a) $\mathbf{I_0}$ maps the (singular) terms of $\mathbf{L_0}$ *onto* **D**.

2. $\mathbf{I_0}$ induces a *valuation function* $\mathbf{V_{M(0)}}$ defined on the set of statements of $\mathbf{L_0}$ as follows:

    (1a(o)) $\mathbf{V_{M(0)}(a = b) = T}$(rue) if and only if $\mathbf{I_0(a)}$ is the same as $\mathbf{I_0(b)}$ for any (singular) terms **a, b** of $\mathbf{L_0}$: otherwise $\mathbf{V_{M(0)}(a = b) = F}$(alse);

    (1b(o)) $\mathbf{V_{M(0)}(P^n(a_1,...,a_n)) = T}$ if and only if $\mathbf{<I_0(a_1),...,I_0(a_n)> \in I_0(P^n)}$, where $\mathbf{a_1,...,a_n}$ are (singular) terms of $\mathbf{L_0}$: otherwise $\mathbf{V_{M(0)}(P^n(a_1,...,a_n)) = F}$;

    (2a(o)) $\mathbf{V_{M(0)}(\sim A) = T}$ if and only if $\mathbf{V_{M(0)}(A) = F}$, and otherwise it $\mathbf{= F}$;

    (2b(o)) $\mathbf{V_{M(0)}(A \& B) = T}$ if and only if $\mathbf{V_{M(0)}(A) = T}$ and $\mathbf{V_{M(0)}(B) = T}$, and otherwise it $\mathbf{= F}$;

    (2c(o)) $\mathbf{V_{M(0)}(A \vee B) = T}$ if and only if $\mathbf{V_{M(0)}(A) = V_{M(0)}(B) = T}$, and otherwise it $\mathbf{= F}$;

    (2d(o)) $\mathbf{V_{M(0)}(A \supset B) = T}$ if and only if $\mathbf{V_{M(0)}(A) = F}$ or $\mathbf{V_{M(0)}(B) = T}$, and otherwise it $\mathbf{= F}$;

(2e(o)) $V_{M(0)}(A \equiv B) = T$ if and only if $V_{M(0)}(A) = V_{M(0)}(B)$,
and otherwise it $= F$;

  (3) $V_{m(0)}(\forall xA) = T$ if and only if $V_{m(0)}(A(a/x))$ for every
(singular) term of L_0, and otherwise it $= F$.

3. To accommodate definite descriptions in **H-BTDD**, first the
interpretation function I_{n+1} is defined as follows:

(1a(n+1)) $I_{n+1}(t) = I_n(t)$ provided t is a (singular) term of L_i
$(i \le n)$;

(1b(n+1)) $I_{n+1}(ixA(x/a)) = d \in D$ such that for some (singular)
term b of L_0, $I_0(b) = d$, and
$V_{M(n)}(A(b/a)) = T$, where $ixA(x/a)$ is a (singular)
term of L_{n+1};[6]

(2(n+1)) $I_{n+1}(P^n) = I_0(P^n)$;

4. I_n induces a valuation function $V_{M(n)}$ defined on the statements
of $L_i (i \le n)$ as follows:

(1a(n)) $V_{M(n)}(s = t) = T$ if and only if
$I_n(s)$ is the same as $I_n(t)$, and otherwise it $= F$;

(1b(n)) $V_{M(n)}(P^n(a_1,...,a_n)) = T$ if and only if
$<I_n(s_1),...,I_n(s_n)> \in I_0(P^n)$, and otherwise it $= F$;

The cases **(2a(n))–(2e(n))** for the connectives are similar to the cases
of **(1a(o))–(1e(o))** above; the universal quantifier is treated thus:

(3n) $V_{M(n)}(\forall xA) = T$ if and only if $V_{M(n)}(A(a/x)) = T$ for every
(singular) term a of L_0, and otherwise it $= F$.

5. The definitions of *logical truth* and *validity* (relative to L_n)
are the conventional ones, and the proof theory **H-BTDD** is
easily shown to be sound in the semantics for **H-BTDD** just
outlined.

4. ASSESSMENT OF H-BTDD

The Hilbert–Bernays informal treatment in the first edition of *Die
Grundlagen* is not hierarchical (as in the formulation above), and indeed, if not reconstructed, is potentially circular. In place of **MA7** they
have an inference rule, the **i**-rule, such that any instance of **MA7** is

---

[6] This clause ensures that each term of the form $ixA(x/a)$ is assigned the unique element
d of **D** that satisfies the basis $A(x/a)$. Its adequacy is guaranteed by condition **(3a)** in
the definition of a model.

derivable given the provability of the appropriate uniqueness condition. In effect, the **i**-rule says that if

$$\exists y(\forall x(A \equiv x = y))$$

is provable, then

$$A(ixA(x/t)/t)$$

is provable. But how can this rule be used to define the syntax of their object language? To what language does the premise of the inference rule belong if the formation rules of the language being defined appeal to that inference rule? Evidently, if the conclusion is in language L_{n+1} the premises of the **i**-rule are tacitly restricted to a sub-language like L_n, a restriction made explicit above, and, hence, an appropriate version of the **i**-rule is derivable in the preceding treatment. The proof is as follows. Suppose the existence and uniqueness conditions *vis-à-vis* $ixA(x/t)$ are provable in L_n. That is, suppose

$$\exists y(\forall x(A \equiv x = y))$$

is provable in L_n. Then, by the formation rules of **H-BTDD**, $ixA(x/t)$ is a singular term of L_{n+1}, $A(ixA(x/t)/t)$ is a statement, and by **MA7**, is also provable therein.

Most importantly, the current formulation of **H-BTDD** is deductively equivalent to a formulation in which **MA7** is replaced by **MD** of Chapter 3. The two principles are mutually deducible in any L_n in which first one, and then the other, is taken as an axiom schema.

Sometimes **H-BTDD** is given a semantic twist (though not by Hilbert and Bernays themselves). The twist amounts to substituting for 'provable' the word 'true' so that **ixA** qualifies as a singular term provided

$$\exists xA$$

and

$$\exists x(\forall y((A \equiv x = y))$$

are jointly true. Because normally the class of true statements is defined only relative to the class of singular terms and statements, the same problem in formally characterizing **H-BTDD** *vis-à-vis* provability arises *vis-à-vis* truth. Probably the matter could be resolved in a manner similar to the characterization of **H-BTDD** above in terms of a hierarchy of languages and associated semantics.

A certain philosophical picture emerges from the preceding account of **H-BTDD**; one might think of it as a theory of logically proper definite descriptions. The analogy is with Russell's theory of logically proper names. According to Russell, if an expression alleged to be a name has no referent, it would not be a name, in the logical sense, but merely "a noise".[7] Similarly, with **H-BTDD** if

$$\exists x(\forall y(A \equiv y = x))$$

is not provable (or not true), the expression **ixA** is grammatically non-significant, a mere noise. One would then have to distinguish between (informal) grammatically proper definite descriptions and logically proper definite descriptions; 'The detective who lived at 221 B Baker Street', for example, would qualify as an example of the former but not of the latter. And one might then proceed to think *à la* Russell about how to paraphrase truths and falsehoods containing merely grammatically proper definite descriptions into the formal language without using logically proper definite descriptions. But this enterprise has limited interest in light of the following complaints.

First, since there is no finite procedure for determining whether a statement is provable, the logical grammar, as noted earlier, is not decidable. Additionally, when attention is turned to the semantic variation of **H-BTDD**, its very grammar would have to await the outcome of the facts; for example, whether the expression

ix(x is a woman who ran the 100 meters in 10.46 seconds in 1988)

qualifies as a singular term depends on whether it is a fact that in 1988 one and only one woman ran the 100 meters in 10.46 seconds. But these grammatical conditions seem quite unnatural.[8] Second, there

---

[7] See *Bertrand Russell: Logic and Knowledge.* (ed. Robert Marsh), Allen Unwin, London (1956), p. 187.

[8] Both of these points have been emphasized by Rudolf Carnap in *Meaning and Necessity*, University of Chicago Press: Chicago, 1947, p. 34. Charity requires that Carnap's undecidability complaint be interpreted as a criticism that the Hilbert-Bernays proposal is too far-reaching, that certain classes of statements clearly do count as grammatical – for example, 'The number identical with 1/0 does not exist' – despite failure of the Hilbert-Bernays provability condition. Undecidability *simpliciter* is no sin, as normal arithmetic makes clear.

is the complaint that mathematicians often make singular existence claims such as 'The null set exists but the Russell set does not'. However, a natural paraphrase of this statement in, for example, the set theory **NF**, or its progeny, as

$$\exists z(z = \emptyset) \;\&\; \sim \exists z(z = ix(\forall y(y \in x \equiv \;\sim (x \in x))))$$

is not provably unique in the right-hand conjunct. Third, Scott has complained that there is another respect in which **H-BTDD** does not conform to actual mathematical practice, as Hilbert and Bernays intended, because definite descriptions are often introduced *before* the existence and uniqueness of their bases are established.[9]

Attempts to overcome the above deficiencies in the original Hilbert and Bernays approach while maintaining their goal to emulate "actual linguistic usage and especially the procedure in mathematics"[10] have been made in recent times by various scholars, two of the most notable being Sören Stenlund and Abraham Robinson.[11] Both Stenlund and Robinson ascribe the undecidability "problem" in **H-BTDD** to a failure to distinguish between two different senses of (singular) 'term', a distinction claimed to be commonplace in modern technical work, but not so when Hilbert and Bernays originally composed their great opus. There is both a "syntactical sense" and a "semantical sense" of 'singular term'. Thus Robinson writes of the Hilbert and Bernays treatment that

This procedure has been criticized because it implies that a formula involving a description can be regarded as well-formed only if it has been proved that the description in question has a denotation. However, this objection is met if – as is natural in a model theoretic approach – we distinguish clearly between well-formedness on the one hand and interpretability in a structure on the other hand.[12]

A similar sentiment is expressed by Stenlund.[13] In other words 'term' in the syntactical sense refers to a certain grammatically significant category whereas 'term' in the semantical sense has to do with whether an expression has a denotation or reference. If not a singular term in

---

[9] Dana Scott, 'Existence and description in formal logic', in *Bertrand Russell; Philosopher of the Century*. (ed. R. Schoenman), Atlantic-Little. Brown, London (1967).

[10] Op. cit., *Die Grundlagen der Mathematik* (first edition) (1968), p. 393.

[11] See footnote 4 for references.

[12] Op. cit., 'Constrained denotation' (1974), p. 493.

[13] Op. cit., *The Logic of Description and Existence*, p. 8.

the semantical sense, an expression is not thereby ruled out as logically ungrammatical, as not being a term in the syntactical sense. When this distinction is made, both Stenlund and Robinson interpret the real message of Hilbert and Bernays to be that it is only in the semantic sense of 'singular term' whether a definite description so qualifies depends on the provability of the existence and uniqueness of its basis. The immediate result in both cases is a vastly simplified characterization of the language of **H-BTDD**. In what follows Stenlund's treatment (hereafter **STDD**) will be the main focus of attention because of its accessibility and the fullness of its discussion.

5. A NEO-HBTDD

Stenlund's **STDD** is really an extension of the original system of Hilbert and Bernays in the sense that the class of provable statements – what Stenlund calls "formulas" – is greater than in **H-BTDD**. For instance, in Stenlund's system

(i) $\exists x(\forall y(P(y) \equiv y = x)) \supset P(ixP(x))$

(in effect) is a provable statement even though

$\exists x(\forall y(P(y) \equiv y = x))$

may itself not be provable. Second, Stenlund's development is a system of natural deduction rather than an axiomatic development. This enables him to display where and when assumptions about the referents of singular terms enter into reasoning while at the same time restricting the class of theorems to truth-valued statements. This is effected by certain conditions on the discharging of assumptions in pieces of reasoning. Third, in contrast to Hilbert and Bernays, Stenlund recognizes truth-valueless statements (among what he calls "formula-expressions"). These may occur as lines in reasoning but never among the provable statements. For instance

(ii) **PixP(x) ⊃ PixP(x)**

is not provable in Stenlund 's system. Moreover, it is truth-valueless if **ixP(x)** fails to refer.

Turning to an outline of the details of Stenlund 's system **STDD** the *vocabulary* is like that of **NTDD** of Chapter 3 except that

~

is eliminated in favor of

$$\rightarrow,$$

and the additional symbol

$$\perp$$

(absurdity). With absurdity available, '$\sim$' can be relegated to a defined sign via the definition

$\sim$**A** for **A** $\rightarrow \perp$.

Though Stenlund introduces function symbols and defines individual constants (names) as 0-placed function symbols, no function symbols will appear here. For the purposes of the current discussion no essential distortion of Stenlund's ideas thereby results.

The **formation rules** are as follows:

(a) Variables **x, y, z, . . .** are *term-expressions*;

(b) Names **a, b, c, . . .** are *term-expressions*;

(c) Any expression of the forms

$$\mathbf{P}t_1, \ldots, t_n$$

and

$$t_1 = t_n$$

are *formula-expressions*;

(d) $\perp$ is a *formula-expression*, and so are

(A & B)

and

(A $\rightarrow$ B)

if **A** and **B** are;

(e) If **A** is a *formula-expression* and **x** is a variable, then

$\forall$**xA** is a *formula-expression*

and

ixA is a *term-expression*;

(f) Nothing else is a *term-expression* or a *formula-expression*.

This definition calls for several comments. First, like the terms and statements of **NTDD** of the previous chapter, the term-expressions and formula-expressions of **STDD** are decidable. Hence, in contrast to Hilbert and Bernays,

$$ix(x \neq x)$$

counts as grammatical in **STDD** even though

$$\exists x(\forall y(y \neq x \equiv y = x)$$

is not provable. Second, 'term-expression' includes cases such as

$$x,$$

$ix(x = a)$, where **a** is a (logically proper) name

and

$$ix(x \neq x).$$

In **NTDD** only the two definite descriptions in the above list count as singular terms; so 'term-expression' in **STDD** does not correspond exactly to 'singular term' in **NTDD**. Indeed, in **STDD** the bottommost definite description in the above list, though a term-expression, does not count as a term at all because it does not refer. A similar asymmetry occurs between Stenlund's use of 'formula-expression' and the statements of **NTDD**. Because

$$x \text{ rotates}$$

is a formula-expression in **STDD**, one cannot pair Stenlund's formula-expressions with the statements of **NTDD**. For example,

$$ix(x \neq x) = ix(x \neq x)$$

though a (true) statement in **NTDD** is merely a formula expression in **STDD** because it is truth-valueless. Third, a simple (or atomic) formula-expression of **STDD** can fail to have a truth-value for one of two reasons. First, it can have a free variable in I – be a quasi-statement in the sense of **NTDD** – hence not purport to be true or false because variables, for Stenlund, do not merely fail to denote but don't even purport to denote.[14] Formula-expressions with free variables are

---

[14] Ibid., p. 37.

merely tools "for making assumptions and, when [the variables] are bound, for expressing propositions". Second, it can contain a term-expression that purports to refer but doesn't and hence, in many cases, cannot have a truth-value.

Many of the differences between **NTDD** and **STDD** would also obtain between **H-BTDD** and **STDD** because the former does not recognize singular terms without existential import or truth-valueless statements. One could, of course, seek to amend **H-BTDD** in the spirit of Stenlund but be more conventional with respect to policies on singular terms and truth-values. In some respects, this may be Robinson's course.[15]

The definitions of *free* and *bound* variables, of the connectives other than

$\sim$

and of the quantifiers, are conventional. $A(t_1,...,t_n)$ means that $t_1,...,t_n$ have been substituted simultaneously (if at all) into $A(x_1,...,x_n)$ under the usual restrictions.

The more distinctive features of **STDD** occur in its **transformation rules**. To this end, Stenlund introduces three meta-logical symbols

I, F, and $\in$,

and *derivations* are treated as tableaus containing expressions of the form

$t \in I, A \in F$, and **A,**

where **t** is a term-expression and **A** is a formula expression.

$t \in I$

means

t denotes

where **t** is not a variable; when **t** is a variable it means something like 't is an arbitrary individual'.[16]

$A \in F$

[15] Op. cit., 'Constrained Denotation', 1978.
[16] Op. cit., *The Logic of Description and Existence*, p. 34.

means

A is true or false,

where **A** is a formula-expression.

A *derivation* is a set of lines each one of which is an *assumption* or follows from other lines in the list in accordance with the *inference rules*. If **A** is the last line of a derivation, **A** is said to be *proved*; if **A** is the last line of a derivation in which all the assumptions have been *discharged*, **A** is said to be a *theorem*.

A derivation is *closed* if all of its premises are discharged; otherwise it is *open*. In **STDD** there is a derivation of **A** only if **A** is true or false. So all theorems – the result of closed derivations – are true or false. As noted earlier this does not mean that the theorems of **STDD** can include only terms (in the semantical sense); for example, definite descriptions with non-unique bases can occur in some theorems of **STDD**.

The code for discharging assumptions follows the method of Prawitz. An assumption is discharged in **STDD** when it is enclosed in brackets, as, for example, in Stenlund's (∀-**rule**):

(∀-**rule**)   [**x** ∈ **I**]

$$\cdot$$
$$\cdot$$
$$\cdot$$

$\mathbf{A(x)} \in \mathbf{F}$

$\overline{\phantom{\mathbf{A(x)} \in \mathbf{F}}}$

∀x**A**(**x**) ∈ **F**,

where **x** does not occur free in any assumption in the derivation of **A**(**x**) ∈ **F** other than in the discharged assumption (in brackets).

The transformation rules fall into three groups: The **I** or *term-introduction rules*, the **F** or *formula-introduction rules*, and the *introduction-elimination rules* for the logical symbols. The complete list in the third category will not be presented here but only the key representatives.

The **I-rules** indicate when an expression is a term in the semantical sense, that is, when it has a referent. They are

(**n–rule**)   **a** ∈ **I**, where **a** is a proper name

and

(i-rule) $\exists x(\forall y(A(y) \equiv y = x))$

$\quad\quad\quad ixA \in I$

The first **I** rule declares it to be axiomatic that any name (individual constant) refers. The second **I** rule says it can be inferred from an assumption establishing the existence and uniqueness of **A** that a definite description, **ixA**, has a referent.

The **F-rules** indicate the conditions under which simple and complex formula-expressions are formulas, that is, have a truth-value. Along with the **(∀-rule)** explained above, they are

(⊥-rule) $\perp \in F$;

   (P-rules) $t_1 \in I,..., t_n \in I$

$Pt_1,...,t_n$;

(I-rule) $t_1 \in I, t_2 \in I$

$t_1 = t_2 \in F$;

(→ -rule)   [A]

                              .
                              .
                              .

        $A \in F$ $B \in F$

$\quad A \to B \in F$

The **I(ntroduction)-E(limination) rules** for logical symbols include

(i I) $\exists x(\forall y(A(y) \equiv y = x)$

$\quad A(ixA(x))$;

(∼ E)   [∼A]

                              .
                              .
                              .

        $\sim A \in F \quad \perp$

A;

(C I) [A]

.

.

.

A ∈ F   B
_____
A → B;

(id I)   t ∈ I
_____
t = t;

(id E)   t = s   A(t)
_____
A(s);

(∀ I)   [x ∈ I]

.

.

.

A(x)
_____
∀xA(x);

(∀ E)   ∀xA(x) t ∈ I
_____
A(t)

The *semantics* of **STDD** are unique. A *model structure* is a triple consisting of a possibly empty set of individuals **D**, an assignment to each individual constant (name) of a member of **D**, and an assignment to an n-adic predicate **P** of a set of n-tuples of members of **D**. Moreover, each individual of **D** has exactly one name, e.g., **a** is the name of **a'** ∈ **D**. To insure this, additional names, e.g., a_1, are added to the language, and hence also axioms of the form

$a_1 \in I.$

The notions of term-expression, formula-expression, etc. are to apply to the extended language.

A partial function **V** is defined on the entire set of closed term-expressions and formula-expressions as follows:

(1)   **V(a)** = **a'** ∈ **D,** for all **a**;

(2) $V(ixA(x))$ is defined if and only if $V(A(a))$ is defined for all $a' \in$ D and $\{a':V(A(a))$ is true$\}$ is a unit set. If $V(ixA(x))$ is defined, then $V(ixA(x)) = b'$ if and only if $\{b'\} = \{a': V(A(a))$ is true$\}$;

(3) $V(Pt_1,...,t_n)$ is defined only if all of $V(t_1),...,V(t_n)$ are defined. If the proviso is satisfied, then $V(Pt_1,...,t_n)$ is $T(rue)$ if and only if $<V(t_1),...,V(t_n)>$ is a member of the set of n-tuples assigned to P; it $= F(alse)$ if $<V(t_1),...,V(t_n)>$ is a member of the complement of the set assigned to P;

(4) $V(t_1 = t_2)$ is defined if and only if both $V(t_1)$ and $V(t_2)$ are defined. Then, if this is the case, $V(t_1 = t_2)$ is T if $V(t_1)$ is the same as $V(t_2)$, and otherwise it $= F$;

(5) $V(\perp) = F$;

(6) $V(A \; \& \; B)$ is defined if and only if both $V(A)$ and $V(B)$ are defined. If $V(A \; \& \; B)$ is defined, then it $= T$ as usual, and otherwise, if defined, $= F$;

(7) $V(A \rightarrow B)$ is defined if and only if both $V(A)$ is defined, and $V(A) = T$ if $V(B)$ is defined. If $V(A \rightarrow B)$ is defined, then it $=$ $T \; (F)$ as usual;

Clauses (2), (3) and (7) are crucial to the non-logical truth of (ii) above. For if $ixP(x)$ is undefined (by (2)), then so is $P(ixP(x))$ by (3), and hence (ii) by (7).

(8) $V(\forall xA(x))$ is defined just in case $V(A(a))$ is defined for all $a \in$ I. If it is defined, then it $= T$ if $V(A(a)) = T$ for all $a' \in D$, and otherwise it $= F$.

A *closed term* t has a referent if $V(t)$ is defined, and a *closed formula-expression* A is T or F if $V(A)$ is defined. STDD is provably *adequate* in the sense that A is a theorem if and only if $A = T$ in all models.[17]

6. ASSESSMENT OF STDD

There can be no question that this neo-Hilbert-Bernays treatment of definite descriptions is vastly simpler than the original approach. Nor can there be any doubt that it is immune to the particular objections that 'The Russell set does not exist' cannot find expression in STDD

[17] Ibid., pp. 66–73.

or that it presumes that mathematicians don't use definite descriptions before their existence and uniqueness assumptions are satisfied. Indeed, it seems to occupy an intermediate position between classical treatments such as Russell's and Frege's, on the one hand, and theories in the free logic tradition, on the other, despite Stenlund's claim that it qualifies as a species of free logic.[18]

Several unique features of Stenlund's **STDD** bear closer scrutiny. First, and leastmost, is Stenlund's sharp contrast between (free) variables and names; the latter refer but the former have no such function. In his semantics free variables play no role; they have no assignments and do not enter into the notion of a model structure. Variables, as noted earlier, are regarded simply as tools, as heuristic devices in certain derivational steps. Nevertheless his discussion of the role of variables is puzzling because the prototype use of free variables is their use in instances of the (∀ **I-rule**), his version of the rule of universal generalization. Thus he writes:

... what we symbolize by writing

$\mathbf{x} \in \mathbf{I}$

.

.

.

$A(\mathbf{x})$

... [is] an argument [that] might begin like this:
Suppose that **x** is an arbitrary natural number ..., then ..., hence $A(\mathbf{x})$[19]

Here **x** evidently *does* function as a referring device, and seems to refer to an arbitrary object, in this case an arbitrary number, the sort of object lately championed by Kit Fine.[20] If so, then the distinction Stenlund wants to make between (free) variables and names does not lie merely in the fact that the former, but not the latter, are merely tools. Rather it lies in the fact that they are both referring devices but refer to different things; names refer to individuals and variables refer

---

[18] Ibid., p. 30.
[19] Ibid., p. 34.
[20] Fine, K., *Reasoning With Arbitrary Objects*. Blackwell, Oxford (1985).

to arbitrary objects. So the syntax, which allows

$$x \in I,$$

is misleading because though informally it says

$$x \text{ refers,}$$

semantically it is not so understood. Stenlund's formal model theory does not always adequately capture his informal intentions.

Second, Stenlund, in contrast to Hilbert and Bernays, rejects the characteristic meta-axiom of **NTDD**. In particular

$$\forall x(x = iyA(y) \equiv \forall y(A(y) \equiv y = x)),$$

fails, when **iyA(y)** is undefined (lacks a reference) as is evident from the clauses for = and → in the semantics for **STDD**. Instead the characteristic meta-axiom of **NTDD** holds in **STDD** only on the condition that

$$\exists x(\forall y(A(y) \equiv y=x)).$$

STDD is more in the spirit of treatments (like Frege) that *amend* **NTDD** rather than in the spirit of treatments (like Hilbert and Bernays) that regards putative counter-examples to **NTDD** as *ungrammatical and hence not genuine.*[21] But this fact also puts **STDD** outside the class of theories of definite descriptions called *free theories of definite descriptions* because the characteristic meta-axiom **MD** of **NTDD** holds in *all* such theories. This observation leads directly to a third and knotty issue *vis a vis* the relationship between **STDD** and logics without existence assumptions.

Stenlund claims that **STDD** qualifies as a species of free logic, or at least as a logic without existence assumptions a la Hintikka. He writes:

Since we also permit terms with no references to occur both in our language and in derivations the deductive system [**STDD**]... can be said to be a 'quantification theory without existential presuppositions (cf. Hintikka, 1959)', or, rather, a deductive system where all existence assumptions are made explicit.[22]

---

[21] See Section 3 of the previous chapter.
[22] Ibid., p. 30.

To be sure, the rules of inference of **STDD** make clear where assumptions about the references of singular term-like expressions enter into derivations. Hintikka's position, however, and indeed the positions of all of the founders of free logic (for instance, Leonard, Leblanc, Schock and yours truly), concern the provable statements and hence, given semantic adequacy, the logical truths of the formal theory.[23] The idea in free logic is that the logical truths assume nothing about the existential import of the singular terms of the language. Free logic is simply an extension of the attitude in modern predicate logic toward general terms (predicates) in logical truths, namely, that logical truths assume nothing at all about whether such expressions are true of any (existing) objects in the extensions of those expressions. Yet Stenlund's **STDD** seems to violate at least the spirit of "logics without existence assumptions" when he declares that

'On our interpretation, the truth or falsity of $[\exists y(t = y)]$ presupposes the truth of "t exists", i.e., of "$[t \in I]$" '[24]

but nevertheless counts

$$\exists y(t = y)$$

logically true when **t** is a name. Clearly it would be truth-valueless were **t** qua name not to have existential import, a possibility excluded in **STDD**. The spirit of free logic is conveyed by the following condition. Consider any class of closed term-like expressions in the sense of Stenlund. Then a logic is free with respect to its term-like expressions just in case no logical truth assumes that the members of any such class refers to an (existent) object. **STDD** violates this condition as the previous discussion shows. Moreover, whether a logic is free or not does not depend upon its being bivalent. For example, Bas van Fraassen has developed a free logic in which truth-value gaps are allowed.[25]

On the other hand, Stenlund's position *vis-à-vis* logics without existence assumptions is puzzling because he appears to want to drive a wedge between his development and free logics. For instance, he

---

[23] See Chapter 8.

[24] Ibid., p. 42.

[25] Bas van Fraassen, 'Singular terms, truthvalue gaps and free logic', *Journal of Philosophy*. 67 (1966), pp. 481–495.

notes that some logicians want to have names that do not refer in the object language, a characteristic of all free logics that have names at all. He regards such a policy as philosophically unsound. This leads directly to the fourth point of issue in **STDD** – the treatment of singular existence statements such as 'Nixon exists', 'the only deposed U.S. president exists' and 'Vulcan exists'.

According to Stenlund all singular existence statements are to be (informally) understood as meta-logical statements about whether the constituent closed term-like expression in the existence statement refers. So, for example, 'Nixon exists' and 'the only deposed U.S. president exists' are to be formally paraphrased or "defined", respectively, as

'Nixon' refers,

that is,

Nixon $\in$ **I**

and

'The only deposed U.S. President' refers,

that is,

The only deposed U.S. President $\in$ **I**.

He says, also, that 'The present King of France does not exist', "interpreted" as

The present king of France $\notin$ **I**,

can nevertheless be "translated" as

$\exists x(\forall y(\mathbf{P}(y) \ y = x))$,

where **P** is the predicate 'present King of France', in virtue of the meta-theorem that

t $\in$ I if and only if I(t) is derivable

where t is a closed term-like expression and **I(t)** is a definite description free formula expression of **STDD**. Despite natural expectations, however, even though it is the case that 'Nixon does not exist', containing the putative (logically proper) name 'Nixon', is "interpreted" as

Nixon $\notin$ **I**

it can nevertheless not be "translated" as

$\sim \exists x(x = \text{Nixon})$

because its translation not only is not provable in **STDD** but would be logically false (a case, perhaps where logical proclivity verges on political overkill). 'The present King of France does not exist', on the other hand, is (informally) true because its translation, though not provable in **STDD**, is not logically false. (There are exceptions, however, for example,

ix(x = t),

where 't' is a name *à la* Stenlund.)

What then of *grammatically proper* names such as 'Nixon' and 'Vulcan' in statements such as 'Nixon exists' and 'Vulcan exists'? If one is engaged in talk about "beings of flesh and blood", Stenlund's choice is to "nixonize" and "vulcanize" away all grammatically proper names a la Quine. Thus 'Nixon exists' becomes

$\exists x(\text{nixonizes}(x)),$

and hence true in agreement with normal intuition while 'Vulcan exists' becomes

$\exists x(\text{vulcanizes }(x)),$

and hence false in agreement with normal intuition.

Stenlund attacks the treatment of singular existence claims in free logic, a position based on the theorem of free logic that

t exists if and only if $\exists x(x = t)$, where **t** is a closed singular term.

First proved by Hintikka in his own version of positive free logic in 1959, it was subsequently extended by me to all free logics, positive, negative or otherwise (See Chapter 2). Stenlund says that the definition

E(xists)!(t) = df $\exists x(x = t)$

involves a confusion of use and mention, the **t** in the definiendum merely being mentioned but in the definiens being used. But this confusion is peculiar to those systems like Stenlund's – for example, the supervaluationally inspired system of Skyrms in 1968 –[26] that regard statements of the form

   E!(t)

as (informal) shorthand for the meta-logical statement

   t ∈ **I**

that is,

   t refers.

This criticism is unsubstantial. It simply is an artifact of Stenlund's development and does not apply to the usual developments of free logic that eschew axioms such as

   t ∈ **I,**

as even the most cursory examination of such systems will reveal.

There is an uncomfortable asymmetry is Stenlund's treatment of statements of the forms

   ... exists

where '...' is filled by a singular term or general term (predicate). In Stenlund's development, and, indeed, in most if not all versions of modern predicate logic, classical or free,

   U.S. presidents exist

and

   kings of France in the 1980s do not exist go
directly into the object language, respectively, as

   ∃**x**(**x** is a U.S. president)

and

   ∼∃**x**(**x** is a king of France in 1980).

They are about, respectively, U.S. presidents and kings of France in 1980, not their common labels, and hence not about whether the

---

[26] Brian Skryms, 'Supervaluations: identity, existence and individual concepts', *Journal of Philosophy*, 16 (1968), pp. 477–483.

general terms (or predicates) 'is a U.S. president' and 'is a king of France in 1980' are true of any existent objects. If Stenlund's arguments in favor of his treatment of singular existence statements were compelling, one would expect that the same strategy would apply to any class of terms, in the absence of any argument to the contrary.

Stenlund asserts that logical systems that accommodate empty names in the object language are mistaken; from "a logical point of view there are no empty names" in the sense that a name (in the logical sense) is "that which names." In the first place, what the logical sense of 'name' is varies from technical development to technical development. Russell's logical sense of 'name' – to which Stenlund's own use seems similar – is very different from Quine's logical sense of 'name'. To say the latter use isn't really logical needs good reasons; otherwise the whole issue reduces to mere stipulation. In the second place, Hilbert and Bernays, given the parallel with Russell's doctrine of (logically) proper names noted earlier, might equally reply that a definite description in the logical sense is that which describes definitely. **STDD** then would be mistaken in the same sense in which Stenlund complains free logics are mistaken. Expressions purporting to be definite descriptions in the logical sense would, if constituted of unprovably unique bases, be "mere noises". In the third place, nothing between Stenlund and the free logician really turns on names; a free logic might consist solely of definite descriptions. Then the issue becomes how best to treat statements containing unfulfilled definite descriptions and especially existence contexts containing such. Stenlund complains that treating 'exists' in 'The only deposed president of the U.S. exists' as a primitive or definable predicate in the object language, is the product confusing the use of that definite description with its mention. But as noted earlier, no such general complaint has any merit in the absence of a specified development. In the fourth place, Stenlund complains that, in some developments of free logic, introduction of an existence predicate in the object language seems to commit one to nonexistent objects and, he says, this is "nonsense."[27] His claim about ontological commitment may be true about some free logics depending on the intended applications of the model structures of such logics, but that such an approach is nonsense

[27] Ibid., p. 40.

is itself nonsense in the light of consistent and complete developments of theories of nonexistent objects in the last few decades.[28]

Finally, a polemical point of some force can be made against Stenlund's treatment of singular existence. Robinson, in his own neo-Hilbert-Bernays development,[29] contemplates a system in which a predicate – call it Φ – is added to the object language and is such that

X refers if and only if Φ (t)

where **X** is a variable ranging over names of **t**. Such a predicate obviously does the job of an existence predicate. In fact this 1972 version of the neo-Hilbert-Bernays theory of definite descriptions by Robinson is very much like many earlier developed versions of free logic.

The upshot of these remarks is that Stenlund's **STDD** represents an intermediate approach between the classical Hilbert-Bernays theory of definite descriptions and free theories of definite descriptions, an approach which though interesting in its own right is not absent of exaggerations or idiosyncrasies.

To sum up, it is clear that the original theory of Hilbert and Bernays does not meet with the original objective of providing a treatment according with mathematical practice. But the neo-Hilbert-Bernays approach of Stenlund also fails to accord with mathematical practice, especially in the matter of singular existence statements such as

1 ÷ 0 does not exist

and

1 × 0 does exist

where the grammatical subjects are treated as complex names. Apparently, only when Hilbert-Bernays is modified in the extreme sense suggested by Robinson, a suggestion amounting to a free theory of definite descriptions, does the fit between actual mathematical practice and the theory of definite descriptions succeed.

---

[28] See, for example, Terence Parsons, *Nonexistent Objects*. Yale, New (1980).
[29] Op. cit., 'Constrained denotation' (1978).

5

Foundations of the Hierarchy of Positive Free Definite Description Theories*

1. INTRODUCTION

From its inception, probably in Frege[1], there has been almost universal agreement on this much about the semantics of definite descriptions – phrases of the form

the **x** which is **F**

where 'the' is used in the singular. When there is exactly one thing **a** that is **F**, then that phrase refers to **a**.[2] Similarly, on the axiomatic side,

---

* Theories in the hierarchy of positive free theories of definite descriptions exemplify different views about which atomic statements containing unfulfilled (improper) definite descriptions are true. (There is no such hierarchy for negative free logics because in such logics every atomic statement containing at least one unfulfilled definite description is false.) Bas van Fraassen, in lectures at Yale in the 1960s, speculated that the hierarchy of positive free definite description theories is linearly ordered by inclusion. However, his conjecture proved to be incorrect, and, in fact, the actual structure is considerably more complex. In this chapter, a description of the languages in which the present study of the hierarchy will be pursued and their semantic interpretations are laid out. Axiomatic theories are then presented and a general strategy for proving completeness is described and applied to the most basic of these theories.

[1] Gottlob Frege, *The Basic Laws of Arithmetic*, (trans. and ed., Montgomery Furth), University of California Press, Berkeley and Los Angeles (1964).

[2] The exception, of course, is Russell who does not regard phrases of the form in question as terms at all, let alone singular terms. Nevertheless, he goes to great lengths to show that the system in *Principia* behaves *as if* this semantic condition were true. See Chapter 1.

69

the basic principle that if **exactly one thing** is **F**, then **the x** that is **F** is
F is accepted universally. In symbols this may rendered as

(1) $\exists!x(F(x)) \supset F((ix)(F(x)))$.

(1) completely determines the logical behavior of fulfilled (proper)
definite descriptions (at least, in extensional contexts). So disagreements are limited to the unfulfilled (improper) case, the case in which
there is no **F** or more than one **F**.

What are the constraints on a theory of unfulfilled (improper) definite descriptions? They are both formal and informal. Formal constraints are those imposed by the underlying logic of quantification;
informal constraints are those which arise from the use of such expressions in informal mathematical, logical and philosophical practice.
Each of these is considered at appropriate points herein.

Definite description theory arose in the context of classical quantification theory, the salient characteristic of which is the principle of
Specification

(2) $(x)P \supset P(t/x)$ for every singular term **t**,

with the corresponding requirement, given the usual semantics, that
every term **t** denotes something in the range of the quantifiers. The
object language counterpart of this last condition is expressed directly
by the following consequence of (2) in classical identity theory:

(3) $(\exists x)(x = t)$.

Evidently one is forced either to deny that definite descriptions are
genuine singular terms, so that (1) does not apply to them, or to provide in some conventional way a reference for unfulfilled descriptions.
Russell took the former course of action; Frege, and after him Carnap,
took the latter.

It is traditional to say, at least as far as mathematical practice goes,
that there are no significant informal constraints on the treatment of
unfulfilled definite descriptions. J. Barkley Rosser, sensitive as he is to
mathematical practice, says the following:

in case $\sim(E_1x)\ F(x)$, we shall simply consider $\iota x\ F(x)$ as a meaningless formula. . . . If the reader is unhappy about using formulas without meaning, he
can arrange for $\iota x\ P$ to have a meaning, regardless of what x and P are, by

adopting the following convention. Choose some arbitrary, fixed object, say the number Π, and agree that...if $\sim(E_1x)$ P, then ιx P is a name of the number Π.[3]

He rejects Russell's proposal strictly on grounds of convenience, and after postulating (1) explicitly for definite descriptions says again that he intends to treat a definite description "as an object" even when it has no meaning. Similarly, Quine refers to unfulfilled definite descriptions as "don't cares" and "waste cases", and says statements containing them can be given any truth-value consistent with logical theory.[4]

Nevertheless, there is counter-pressure to the conventionalist policy. For instance, Frege, after suggesting a "single chosen object" theory, argues for a more discriminating treatment of unfulfilled definite descriptions. He says:

If there were no irrational numbers – as has indeed been maintained – then even the proper name 'The positive square root of 2' would be without a denotation, at least by the straightforward sense of the words, without a special stipulation. And if we were to give this proper name a denotation expressly, then the object denoted would have no connection with the formation of the name, and we should not be entitled to infer that it were a positive square root of 2, while yet we should only be too inclined to conclude just that.[5]

To deal with this problem, Frege introduces into his ideal language a function \ whose value for a singular course of values is the unique object in it, and otherwise is the identity function. The effect (if translation from ordinary language into ideal language is taken into account) is to associate with an unfulfilled definite description the course of values of its basis (matrix), an association that is less conventionalist than the arbitrary chosen object approach.

Rosser, too, seems sensitive on this matter. Despite offering the squeamish reader a single chosen object, it turns out that his axioms are not complete for this interpretation; instead of (in effect) the schema

(4) $(\sim E_1xP \ \& \ \sim E_1xQ) \supset \iota xP = \iota xQ$

[3] J. Barkley Rosser, *Logic for Mathematicians*, McGraw Hill, New York (1953), p. 182.

[4] W. V. Quine, *Word and Object*, Wiley, New York (1960), p. 182.

[5] Op cit., *The Basic Laws of Arithmetic*, p. 50.

he has only the weaker principle

(5) $(\mathbf{x})(\mathbf{P} \equiv \mathbf{Q}) \supset \iota\mathbf{x}\mathbf{P} = \iota\mathbf{x}\mathbf{Q},$

a principle also satisfied by the version of Frege's theory just sketched.

So there is pressure from some quarters in informal practice to allow the basis of an unfulfilled definite description to play a role in determining its referent beyond the simple certification of non-fulfillment. But this in turn requires that more than one object play the role of referent for such definite descriptions. Systems of this sort will be called *extended* chosen object theories. The original Frege proposal, and as it is exemplified in Carnap's Method IIIa,[6] is an extended chosen object theory because, for example, the unfulfilled definite descriptions 'the prime number' and 'the largest positive integer' have different referents, the first being a class of infinitely many objects, and the second, the null class.

(4) and (5) are examples of extensionality conditions on definite descriptions; they can be construed as telling one how much of the meaning of $\mathbf{P}$ is relevant to the determination of the reference of $\mathbf{ix(P)}$, when $\mathbf{P}$ is not true of exactly one object. Variation in such principles is one of the principal dimensions of the space of positive free definite description theories.

(5) is weaker than (4); it brings more of the meaning of $\mathbf{P}$ into play. How far can one go in this direction? Various answers bring out yet further features of the informal use of definite descriptions. In the first place, it is not really the meaning of $\mathbf{P}$ *simpliciter* that is at stake, but rather the meaning of $\mathbf{P}$ *qua* predicate of $\mathbf{x}$; such is the significance of the bound variable $\mathbf{x}$ in $\mathbf{ix}$. So the antecedent of (4) really says that $\mathbf{P}$ and $\mathbf{Q}$ *qua* predicates of $\mathbf{x}$ have the same extension.[7]

---

[6] Rudolf Carnap, *Meaning and Necssity*, University of Chicago Press, Chicago (1947), pp. 35–36.

[7] The shift from the schematc letters $\mathbf{P}$, $\mathbf{Q}$, ... to the schematic letters $\mathbf{A}$, $\mathbf{B}$, ... is done to avoid confusion with other uses the former are put to later in this essay. The term 'extension' as used here goes back to the Port Royal logic (Arnauld, 1662), and means the set of objects of which the predicate ("general idea") is true. It is there contrasted with the "comprehension" of a general idea, the latter defined as "the constituent parts of the idea". This notion has subsequently come to be called the intension of the predicate; the term 'comprehension' will be reserved for a different notion in what follows, a notion more closely resembling C. I. Lewis's well-known use of 'comprehension'.

To facilitate discussion a heuristic device will be introduced; the notation $(\mathbf{x},\mathbf{A})$ is to stand for '$\mathbf{A}$ *qua* predicate of $\mathbf{x}$'. Without, as Frege does, reconstruing the grammatical form of descriptions as $i(\mathbf{x},\mathbf{A})$, one can nevertheless recognize that the reference of $(\mathbf{ix})\mathbf{A}$ is a function of the meaning of $(\mathbf{x},\mathbf{A})$; indeed, definite description theory is the theory of just this function.[8]

The *extension* of $(\mathbf{x},\mathbf{A})$ is the set of things of which it is true. This is clearly part of its meaning, and is all that is required to determine whether $(\mathbf{ix})\mathbf{A}$ is fulfilled or not. Indeed, in the fulfilled case, it is all that is required to determine its referent. But it may be questioned whether the extension of $(\mathbf{x},\mathbf{A})$ in general suffices for determination of reference in the unfulfilled case. For instance, imagine the domain of discourse to be the positive integers. Then the single chosen object theory requires that the square root of 2 equal $1 - 2$, but this is very implausible. The source of this implausibility may be the fact that, by analogy with certain fulfilled cases, one is inclined to say that though $(\sqrt{2}) = 2$ and $3 + (1 - 2) = 2$, neither $(1 - 2)^2 = 2$ nor $3 + \sqrt{2} = 2$. But this difference cannot be explained by a difference in the extensions of $(\mathbf{x},\mathbf{x}^2 = 2)$ and $(\mathbf{x},\mathbf{x} + 2 = 1)$, both of which are empty.

What other part of the meaning is required? One obvious factor is knowledge of that of which the predicates are true in the wider domain of the negative and irrational integers. Another factor, not requiring a change in the initially imagined domain of discourse – the positive integers – is the rules of computation *vis-à-vis* the manipulation of symbols. This latter factor suggests that the aspect of meaning of $(\mathbf{x},\mathbf{A})$ to be taken into account is very close to the expression itself.

So, assume for the moment, that in the unfulfilled case the referent of a description is determined simply by the syntactic form of the definite description, that any two distinct unfulfilled descriptions have distinct values. This would amount to imposing *no* conditions like (4) and (5) at all, a course which was actually followed, though, perhaps inadvertently, by the originators of some of the systems to be analyzed. Such systems, to borrow Cresswell's useful label, are hyper-intensional.[9]

[8] Though the Fregean construal would no doubt contribute to logical purity, it introduces substantial and unnecessary complications into the current study.

[9] Cresswell, however, was concerned with modal logic rather than term operators when he coined the expression. Examples of such systems include the Hilbert-Bernays ∈-calculus and the definite description theory, MFD, of Lambert and van Fraassen.

Precisely at this point another further formal constraint emerges, a constraint bound to influence further reflections on the determinants of the referents of unfulfilled definite descriptions. The underlying classical logic postulates unrestricted substitutivity of identity (in statements) in the form

(6) $s = t \supset (A(s/x) \supset A(t/x))$,

from which substitutivity of identity (in singular terms)

(7) $s = t \supset r(s/x) = r(t/x)$

readily follows. This in turn implies that one can prove, for instance, that

(8) $(\imath x)(x^2 = 2) = (\imath x)(x^2 = (1 + 1))$.

So apparently not only the syntactical shape of $(x, x^2 = 2)$ must be taken into account in determining the referent of an unfulfilled definite description but also the referents of at least some of its parts.

Why not simply give up the constraint of substitutivity, as is done, for instance, in extending classical to modal logic? Because there is another informal reason for retaining this formal constraint, a reason further complicating the semantics of definite descriptions.

Perhaps the most important application of definite descriptions in mathematics is in the explication of function notation, especially in connection with partial functions. Thus Whitehead and Russell say

> The notation "$[(\imath x)(\Phi x)]$"... is seldom used, being chiefly required to lead up to another notation, namely "R'y," meaning 'the object having the relation R to y.' ... R'y is a function of y, but not a propositional function; we shall call it a *descriptive* function. All the ordinary functions of mathematics are of this kind.[10]

From this point of view, when A contains singular terms $t_1, ..., t_n$, $(\imath x)A$ should be regarded as having the form $f(t_1, ..., t_n)$ and in particular, $(\imath x)(x^2 = 2)$ and $(\imath x)(x + 2 = 1)$ as having the forms $f(2)$ and $g(2,1)$, respectively. The role of the definite description is now just to analyze

---

[10] Alfred N. Whitehead and Bertrand Russell, *Principia Mathematica*, (second edition), vol. 1, Cambridge University Press (1950), p. 31.

f and **g**, rather than the singular terms they are used to form. So it appears natural to restrict the part of the meaning of $(\mathbf{x}, \mathbf{x}^2 = 2)$ required for determining its referent to *exclude* the meanings (other than referents, perhaps) of the singular terms (or variables) which occur (free) in it.

A way of representing this "general" (non-singular) part of an expression was developed for proof theoretic purposes by John von Neumann, and later exploited at length in Hilbert and Bernays in 1934.[11] They called it the method of "name forms"; the more descriptive term "skeleton" is used here.[12] To this end a new family of variables, called *placeholders*, $\mathbf{n}_1, \mathbf{n}_2, ...,$ will be introduced, and a skeleton defined as an expression in which the only free variables are placeholders, occurring once each and forming, in left to right order, an initial segment of the above sequence. For any expression **E** of the language there is a unique skeleton from which it may be obtained by substitution as well as the sequence of those substituends. These will be called, respectively, the skeleton (of the expression) **E** (**skel E**) and the arguments of **E**. Thus the skeleton of $(\mathbf{x}, \mathbf{x}^2 = 2)$ is $(\mathbf{x}, \mathbf{x}^2 = \mathbf{n}_1)$, and its argument is (2), while $\mathbf{skel}(\mathbf{x}, \mathbf{x} + 2 = 1)$ is $(\mathbf{x}, \mathbf{x} + \mathbf{n}_1 = \mathbf{n}_2)$ and its arguments are $(\mathbf{2}, \mathbf{1})$.

It must be acknowledged, however, that there are also informal considerations militating against the extended chosen object policy. Thus Russell, in 'On denoting', complains that though Frege's theory

... may not lead to actual logical error [it] is plainly artificial, and does not give an exact analysis of the matter.[13]

Of course, one may argue that Russell's objection does not take into account that Frege's goal was not to give an analysis of the logical form of statements in colloquial discourse, but rather to improve on such discourse for scientific purposes. But then, there is also Rosser,

---

[11] David Hilbert and Paul Bernays, *Grundlagen der Mathematik*, Springer, Berlin (1934, 1939).

[12] The use of skeletons in the semantics of term forming operators probably originates with Asser in providing a semantics for the Hilbert $\in$-calculus. See Günter Asser, 'Theorie der logischen Auswahl funktionen', *Zeitschrift für mathematische Logik und Grundlagen der Mathematik*, 3 (1957), pp. 30–68.

[13] Bertrand Russell, 'On denoting' in *Logic and Knowledge* (ed. R. Marsh), George Allen & Unwin; London (1966), p. 47. This is a reprint of Russell's famous article in 1905 in the journal *Mind*.

who despite actually using the extended chosen object method, admits that the

fact that we allow in the symbolic logic terms with no meaning leads to minor discrepancies between our terminology and that current in mathematics. This comes about as follows.

By Thm.V11.2.2,

$\vdash (f,x)(E_1z).z = f(x)$

This would seem to say that f(x) is defined and unique for every f and x. Such is not the case at all.

The fact that Axiom schemes 8, 9 and 10 give f(x) a sort of spurious existence for any x or f is a trifle confusing. What it means is that 'existence of f(x)' in the mathematical sense is not actually a statement about f(x), despite its misleading grammatical form, but rather a statement about f and x separately. Thus the statement 'f(x) is defined at x' would be rendered symbolically as $(E_1y)xfy$, and in this form does not contain the combination f(x) at all.[14]

So Rosser is driven into a kind of contextualism about "exists", and one gets the impression that if pressed, he would fall back on a Russellian position for all uses of definite descriptions.

Moreover, the argument above (based on a single domain of positive integers) to impugn single chosen object theories can also be used against extended chosen object theories. For whatever integer one may choose as the referent of '$\sqrt{2}$', $(\sqrt{2})^2 = 2$ will turn out false, contrary to the intuition that it should be true. Conversely, Suppes claims:

Many mathematicians would be uneasy at seeing:

$2/0 = 0$

and thus, say:

$1 + 2/0 = 1.$[15] [Suppes 1957, p. 164]

The upshot of these reflections is that if one wishes to retain the plausible view that definite descriptions function grammatically as genuine singular terms, then one has to abandon the classical assumption evident in (3) that every such term denotes something in the

---

[14] Op. cit., *Logic for Mathematicians*, p. 319.
[15] Patrick Suppes, *Introduction to Logic*, Van Nostrand, Princeton (1957), p. 164.

range of the quantifiers.[16] The resulting logic is a *free logic*, and the theories of definite descriptions based on it are *free definite description theories*.

2. POSITIVE FREE DEFINITE DESCRIPTION THEORIES: PRELIMINARY REMARKS

If the classical assumption that every singular term denotes something in the range of the quantifiers is rejected, then such terms either denote something not in the range of the quantifiers, or denote nothing at all. Both sorts of interpretation of free logic are available in the literature.[17] The former course is followed here for reasons of convenience rather than doctrine. The "outer domain" is to be regarded as an "artifact of the model" carrying no true ontological commitment. The question whether it is to be reduced or explained away in some robustly realistic fashion may be resolved differently in different cases; in the above mathematical example, for instance, there seems little objection to admitting negative and irrational numbers, while admitting round squares is much more problematic.[18]

Russell rejected Meinong's theory, which the current policy resembles, as one that inevitably leads to contradiction; for the round square would have to be both round and square, the existent king of France existent and nonexistent – or so he thought. The current approach is not designed to cope with such cases; definite descriptions with "inconsistent" bases (such as the above) will be treated by a non-Fregeian version of the chosen object theory.[19]

[16] Interestingly enough, Rosser himself flirted with such a system in an early paper in 1939, thereby hinting at what appears to be the first "free definite description theory." But, apparently, he never pursued the idea.

[17] For an example of the first kind of development, see Robert K Meyer and Karel Lambert, 'Universally free logic and standard quantification theory', *Journal of Symbolic Logic*, 33 (1968), pp. 8–26. For an example of the second kind of development, see Bas van Fraassen, 'The completeness of free logic', *Zeitschrift für mathematische Logik und Grundlagen der Mathematik*, 12 (1966), pp. 219–234.

[18] Karel Lambert, 'Nonexistent objects: why theories about them is important', *Grazer Philosophische Studien*, 25–26, (1985–86), pp. 439–446.

[19] For an answer to Russell's worries, closer to Meinong in spirit, see Terence Parsons, *Nonexistent Objects*, Yale University Press, New Haven (1980). Parsons' treatment, however, involves extensive (and avoidable) complications in the theory of predication.

On the other hand, a rich variety of theories of "consistent" unfulfilled definite descriptions, differing on the one hand in their extensionality conditions (as depicted above), and on the other in the conditions they require for **F((ix)F(x))** to hold – the counterpart here of Meinong's *Sosein* principle. It is worth noting that the admission of objects not in the range of the quantifiers, though not directly motivated by considerations of extensionality, introduces a new consideration *vis-à-vis* predicates such as **(x, A)**. As in the earlier mathematical example appealing to a domain beyond the positive integers, the set of objects of which **(x, A)** is true in the wider domain becomes an essential object of interest. C. I. Lewis's *comprehension* will be used to refer to this set, though his interpretation of the elements of the outer domain as just the possible objects is not endorsed here.

A quantifier ranging over the wider domain will be introduced; written as **[x]A**, it may be read as 'For all objects **x** such that **A**'. Because judgment on the interpretation of the outer domain has been suspended, quantifying over it should raise no further ontological worries. On the other hand, it allows considerable economy in some of the technical work to follow, since it allows formulation of some axioms that apparently cannot be expressed without it. The logic of this quantifier is precisely that of the well understood first order classical logic of quantifiers.

These preliminary remarks end with an informal prologue to the ensuing semantics of definite descriptions. A definite description of the form **(ix)A** will be regarded as having the *logical* form **i(skel(x, A))(t₁,...,tₖ)**, where **t₁** through **tₖ** are the arguments of **skel(x, A)** as characterized in the previous section. The meaning of a k-ary predicate skeleton is a pair consisting, first, of the skeleton itself and, second, of the comprehension of the skeleton. The comprehension of a skeleton will be taken to be a k-ary function from objects to subsets of the wider domain. If one imagines a predicate skeleton as a *parameterized complex predicate*, then its comprehension is just the function associating with given values of the parameters the comprehension of the complex predicate they determine. So the interpretation of **i** will be a function that associates with the meaning of a k-ary predicate skeleton – that is, the pair consisting of the skeleton and a k-ary function from objects to subsets of the wider domain – a k-ary function from objects to objects.[20] The capital italic *I* is used to designate this function.

These complexities are required to handle the hyper-intensional systems, but in fact in most cases a much simpler interpretation will suffice, one which takes as arguments simply predicate comprehensions or extensions – respectively, subsets of the wider or narrower domain.

3. POSITIVE FREE DEFINITE DESCRIPTION THEORIES: THE FORMAL LANGUAGE

Turning to formalities, some of the theories studied herein are even more general than free definite description theories. These theories, formulated in a number of first order languages, are sub-languages of a comprehensive language *L* now to be defined. The primitive improper symbols of *L* are

$\sim$ & i $\forall$ ( ) [ ]

Of these, the first two represent truth-functional negation and conjunction, respectively, the third is (part of) the description operator, the fourth is part of the two universal quantifiers, the narrower ($\forall$x) and the wider [$\forall$x], respectively.

Primitive proper symbols include denumerably many individual variables, denumerably many function symbols of each finite degree and denumerably many relation symbols of each finite degree, including the singulary

E!

(for existence) and the binary

=

(for identity). The object language will not be exhibited in detail, but **x, y,** etc. are meta-variables for individual variables; **a, b,** etc. for

---

[20] Care must be taken to distinguish between the comprehension of a skeleton and the comprehension of a predicate.

individual constants; **f, g, etc.** for function symbols; and **P, Q, etc.** for relation symbols.

Expressions are finite strings of symbols. The well-formed expressions divide into singular terms and (possibly open) sentences along familiar lines:

[1]   Every individual variable or constant is a *singular term*.
[2]   If **f** is of degree **n** and $t_1...t_n$ are *singular terms*, then $f(t_1...t_n)$ is a *singular term*.
[3]   If **A** is a *sentence*, **(ix)A** is a *singular term*.
[4]   If **P** is of degree **n** and $t_1...t_n$ are *singular terms*, then $P(t_1...t_n)$ is a *sentence*.
[5]   If **A, B** are *sentences*, so are ~**A**, [**A&B**], (**x**)(**A**) and [**x**](**A**).
[6]   If **A** is a sentence, (**x,A**) is an **@***complex predicate*.
[7]   That's all.

Clause **[3]** characterizes **(ix)** grammatically as a variable-binding term operator (vbto)[21]; much of what ensues applies to any such operator. Clause **[6]** introduces a special kind of predicate, the @complex predicates mentioned in the introduction. Note that @complex predicates never occur explicitly as syntactical parts of any expression, though they will be treated here as having implicit occurrences in definite descriptions.

r, s, t, etc. are meta-variables for singular terms; **A, B, etc.** for sentences; **G, H, etc.** for @complex predicates and **E, F, etc.** for arbitrary well-formed expressions. Bound and free occurrences of variables are as usual, with the addition that occurrences of **x** in (**x,A**) are bound; the notion is extended also to arbitrary terms by declaring bound in **E** those occurrences of terms **t** which contain occurrences of variables free in **t** but bound in **E**. Expressions without free variables are *closed* and closed sentences are *statements*. **E̲** is used to designate the class of free variables of **E**, and $E(t_1...t_n/x_1...x_n)$ to designate the result of simultaneous substitution of the t_i for the free occurrences of the

[21] The study of general vbtos apparently was initiated by Hatcher, W., *Foundations of Mathematics*, W. B. Saunders, Philadelphia (1968), pursued in Corcoran, J. and Herring, J., 'Notes on a semantic analysis of variable binding term operators', *Logique et Analyse*, 55 (1972), pp. 644–657, and, again, in Corcoran, J., Hatcher, W., and Herring. J, 'Variable binding term operators', *Zeitschrift f. math. Logik und Grndlangen d. Math.*, 18 (1972), pp. 177–182.

corresponding x_i in **E**. $(t_1...t_n/x_1...x_n)$ stands for the class of expressions **E** in which the indicated substitutions can be made without illicit binding of free variables, i.e., such that for each $i \leq n$, if **x** has a free occurrence in any part **(iy)A**, **(y)A** or **[y]A** of **E**, then **y** is not free in t_i.

Next skeletons are introduced for the purposes mentioned in the introduction. Let the language L_n be the extension of **L** obtained by the addition of denumerably many new "placeholders" n_1, n_2, ... These are treated like individual constants in the syntax of L_n, which otherwise is like that of **L**. A **skeleton** is an expression of L_n whose free terms are, in left to right order, n_1 to n_k, for some **k**. The number **k** is the **degree** of the skeleton. If **F** is a **k-ary skeleton**, and **E** is an expression of **L**, then **F** is a skeleton of **E** if and only if there are terms t_i, ..., t_k such that $F \in (t_1...t_k/n_1...n_k)$ and $E = F(t_1...t_k/n_1...n_k)$. The following can be proved by induction on the structure of **E**:.

Lemma 1: For any set of variables **V**, there is a unique **V-skeleton** of **E**,

where the definition of the latter is like that of '**skeleton**' with '**free term**' amended to '**free term not in V**'.

The **skeleton** of **E** is called **Skel (E)**, and the **degree** of **E is** stipulated to be the same as that of **Skel (E)**. Furthermore, the terms $t_1...t_k$ indicated above are uniquely determined; they are called the **arguments** of **E**. In practice only skeletons of @complex predicates are required; they will be used in specifying the interpretation of the description operator in hyper-"intensional" systems. At one crucial point the notion of the **x-skeleton** of a sentence **A** is needed; it is defined to be the second member of **skel (x,A)**.

The following lemma is central to the use of skeletons; it can again be proved by induction on the structure of the expression **E**:

Lemma 2: If $E \in (t/x)$, then **skel** $(E(t/x)) =$ **skel E**; also t_1 is an argument of **E** if and only if $t_1(t/x)$ is an argument of $E(t/x)$.

The following are adopted as abbreviations; in **D8** and **D9**, **y** is to be the first variable not free in **A** or **x**, while in **D10**, **x** is the first variable not free in **s** or **t**.

> **D1** $[A \supset B]$ for $\sim[A \,\&\, \sim B]$
>
> **D2** $[A \lor B]$ for $\sim A \supset B$
>
> **D3** $[A \equiv B]$ for $[A \supset B] \,\&\, [B \supset A]$

D4	$s = t$	for	$= (s,t)$
D5	$s \neq t$	for	$\sim =(s,t)$
D6	$(\exists x)A$	for	$\sim(x)\sim A$
D7	$[\exists x]A$	for	$\sim[x]\sim A$
D8	$(\exists_1 x)A$	for	$(\exists y)(x)[A \equiv x = y]$
D9	$[\exists_1 x]A$	for	$[\exists y][x][A \equiv x = y]$
D10	$s \approx t$	for	$(x)(s = x \equiv t = x)$

SEMANTICS I: FREE LOGIC

This section presents the interpretation of the language **without** the description operator, that is, the underlying free logic. A **free model structure** is a pair S = <**D1, D2**> of sets with **DI** $\subseteq$ **D2** and **D2** nonempty. A **valuation v** in S assigns to each constant of **L** an element of **D2**, to each k-ary function symbol a **k-ary** function from **D2** to itself, and to each k-ary relation symbol a **k-ary** relation on **D2**. In particular, **v** assigns the identity relation on **D2** to = and **D1** to **E!**. An assignment α maps each individual variable to an element of **D2**. $\alpha(d_1...d_n/x_1...x_n)$ is the assignment like α except for mapping x_i to D_i for $i \leq n$.

A valuation and assignment jointly determine a value in **D2** for each **i-free** singular term and a truth-value in $\{T, F\}$ for each such sentence according to the usual recipe:

[1] $v\alpha c = vc$ for constants **c**, and $v\alpha x = \alpha x$ for variables **x**

[2] $v\alpha f(t_1...t_n) = vf(v\alpha t_1...v\alpha t_n)$

[3] $v\alpha P(t_1...t_n) = T$ iff $<v\alpha t_1...v\alpha t_n> \in vP$

[4] $v\alpha \sim A = T$ iff $v\alpha A = F$

[5] $v\alpha(A\&B) = T$ iff $v\alpha A = v\alpha B = T$

[6] $v\alpha(x)A = T$ iff for all $d \in D1$, $v\alpha(d/x)A = T$

[7] $v\alpha[x]A = T$ iff for all $d \in D2$, $v\alpha(d/x)A = T$.

If **A** is a statement, $v\alpha A$ is independent of α and we write simply **vA**. A **free model** is a pair **M** = <**S, v**>, **S** a **free model structure** and **v** a **valuation** in **S**. A statement **A** is true in **M** = <**S, v**> iff **vA** = **T**; validity and satisfiablility in **M** and unconditional validity and satisfiability can now be defined as usual.

The semantics for **[x]** is precisely that of the classical universal quantifier ranging over **D2**, with **(x)** as a restricted quantifier ranging over a subset of **D1** defined by **E!**. Thus **(x)A** is logically equivalent to **[x]** **(E!(x) ⊃ A)** for every **A**.

5. SEMANTICS II: @COMPLEX PREDICATES AND SKELETONS

The comprehension of a **@complex predicate** is the set of objects, inner or outer, of which it is true; the extension is the restriction of the comprehension to the inner domain. Formally, this is

[8] $v\alpha(x,A) = \{d \in D2: v\alpha(d/x)A = T\}$

The **comprehension** of **(x, A)** **(on v and α)** is just $v\alpha(x, A)$; its **extension** is **D1** $\cap$ $v\alpha(x, A)$.

The use of skeletons in their original proof-theoretic application by Hilbert and Bernays was purely formal, no special interpretation being attached to them. Here they have semantic values. In particular, a **k-ary skeleton** stands for a **k-ary function** from **objects** to entities appropriate to its syntactical category. Thus, a term skeleton will represent a function from objects to objects, a sentence skeleton a function from objects to truth-values, and a predicate skeleton a function from objects to sets of objects. The following definition makes use in the meta-language of the ubiquitous lambda operator of Church to form names of functions.

Let **E** be a **k-ary skeleton**, let $x_1...x_k$ be the first **k** variables of **L** not occurring in **E**, and let $E' = E(x_1...x_k/n_1...n_k)$. Then

[9] $v\alpha E = \lambda d_1...d_k(v\alpha(d_1...d_k/x_1...x_k)E')$

Since the x_i are all the free variables of **E'**, the value of **vαE** will be independent of α, and one can write simply **vE**. **vE** is **the comprehension of E**.

4. SEMANTICS III: DEFINITE DESCRIPTIONS

The ingredients necessary for the interpretation of terms of the form **(ix)A** have now been assembled. Recall that such a term is to be analyzed as a term as of the form $f(t_1...t_k)$, where t_1 etc. are the arguments of the term and **f** is its skeleton. This means that $v\alpha(ix)A$ is defined to be the result of applying the function determined by

skel(ix)A to $v\alpha t_1$, etc. But this function in turn is determined by the meaning of the @complex predicate skeleton **skel(x, A)**. Accordingly, a (higher-order) function I^{22} is introduced that associates with the meaning of **skel(x, A)** a function from objects to objects of the same degree.

How much of the "meaning of **skel(x, A)**" must be simulated to achieve the purposes expressed in the Introduction? Nothing less than the skeleton itself! For it is needed to individuate definite descriptions in the weakest systems (See Section 8 below). On the other hand, the meaning of **skel(x, A)** must carry sufficient semantical information to enable one to determine, for instance, whether or not the definite description is fulfilled, and if so, what its referent will be. For this latter purpose the comprehension **v(skel(x, A))** as defined above will be sufficient. Accordingly, the meaning of **skel(x, A)** on the valuation **v** will be simulated by the pair **(skel(x, A), v(skel(x, A)))**, called the **hyper-intension** of **skel(x, A)**; the specification function *I* will be defined on them. If **skel(x, A)** is **k-ary**, then the **hyper-intension** is **k-ary** as well, and the value of *I* for a **k-ary argument** is a **k-ary function** from objects to objects. So, finally, the list of semantic clauses may be completed as follows:

[10] $v\alpha(\text{ix})A = (I(\text{skel}(x, A), v(\text{skel}(x, A))))(v\alpha(t_1...v\alpha t_k))$.

Because all of the clauses above form part of a recursive definition, that definition in fact produces a unique valuation **v** satisfying the clauses for all assignments given the value of **v** for the primitive proper symbols, given the following understanding of clause **[10]**. In the case of **[10]** the uniqeness of **v** is guaranteed only if *I* is defined for all **hyper-intensions**. Trivially this is the case if *I* is defined for all entities of the logical type "pair consisting of a @complex predicate skeleton and a function from objects to D2" (**possible hyper-intensions**). For present purposes, however, it is convenient to require only that *I* be defined on those possible **hyper-intensions** that are "expressible", that is, those that are *actually* the **hyper-intension** of some skeleton. Therefore, a set **EX** (for 'expressible') of the **possible hyper-intensions** will be added to the apparatus, and this set understood as the domain of definition

---

[22] In the ensuing discussion, this function is sometimes referred to as the *specification function*.

of I. Since expressibility itself involves the valuation **v**, the definition below of a model below must be existential in form.

A **free definite description structure** is a triple $D = <S, I, EX>$, where **S** is a **free model structure** and **I** and **EX** are as above. A **free definite description model** (henceforth simply model) is a pair **<D, v>** such that there exist functions **v** for each assignment α, and $\mathbf{vd_1...d_k}$ for $d_1...d_k$ in **D2**, that satisfy the semantic clauses **[1]–[10]**. Satisfiability and validity are now defined as usual.

Two fundamental lemmas are:

Lemma 3: (**Local determination**) If **L'** is any sub-language of $\mathbf{L_n}$ including **L**, **<S,v>** a model for **L'**, **E** any expression of $\mathbf{L_n}$ that belongs to **L'**, and α and α' assignments agreeing on the free variables of **E**, then $\mathbf{v\alpha E} = \mathbf{v\alpha' E}$.

Lemma 4: (**Semantic substitution**) Under the above hypotheses about **L'**, **<S,v>** and **E**, if $\mathbf{E} \in (t/x)$ then $\mathbf{v\alpha E}(t/x) = \mathbf{v}(\alpha\,(\mathbf{v\alpha}t/x))\mathbf{E}$.[23]

To illustrate the machinery, the case where $\mathbf{E} = \mathbf{(iy)A}$ of **Lemma 4** is proved.

Let **p** be $\alpha(\mathbf{v\alpha}t/x)$. If **x** is not free in **E**, it follows from **Lemma 3**. Suppose it is. Then $\mathbf{E}(t/x) = \mathbf{(iy)}(\mathbf{A}(t/x))$ and $\mathbf{A} \in (t/x)$. Now $\mathbf{v\alpha E}(t/x) = (I(\mathbf{skel}(y, \mathbf{A}(t/x)), \mathbf{v}(\mathbf{skel}(y, \mathbf{A}(t/x)))))(\mathbf{v\alpha}(s_1,...)$, where s_1, etc. are the arguments of $(\mathbf{y}, \mathbf{A}(t/x))$ $(= (\mathbf{y}, \mathbf{A})(t/x))$, **by [10]**. By **Lemma 1**, $\mathbf{skel}((\mathbf{y}, \mathbf{A})(t/x)) = \mathbf{skel}(\mathbf{y}, \mathbf{A})$, and for each **i**, $s_i = t_i(t/x)$ where t_i is the **i-th** argument of $(\mathbf{y}, \mathbf{A})$ and $t_i \in (t/x)$. By inductive hypothesis, $\mathbf{v\alpha}s_i = \mathbf{vpt_i}$ for each **i**. Hence $\mathbf{v\alpha E}(t/x) = (I(\mathbf{skel}(y, \mathbf{A}), \mathbf{v}(\mathbf{skel}(y, \mathbf{A}))))(\mathbf{vpt_1},...) = \mathbf{vpE}$ (by **[10]**) as required.

6. A MINIMAL SYSTEM

The semantics of Sections 3.–5. specify nothing about the nature of the specification function I beyond its logical type; the semantics is intended to be completely general for theories of a variable binding operator based on (positive) free logic.[24] Any set of postulates for

---

[23] **Lemma 4** is stated only for single substitutions, but holds equally for simultaneous substitution.

[24] A general semantics and complete axiom system for **vbtos** in classical logic is presented by Corcoran, Hatcher and Herring 1972. They also make use of skeletons, calling them canonical terms, and provide an extremely elegant completeness proof

classical predicate logic will do for the quantifier **[x]**, together with the following schema relating the inner and outer quantifiers

    [E]   (x)A ≡ [x](E!(x) ⊃ A) is an axiom for every **A**.

For definiteness the following Hilbert inspired formulation is adopted here.

      [T]   Any substitution instance of a **tautology** is an axiom.

      [UI]   **[x]A ⊃ A(t/x)** is an axiom if **t ∈ (t/x)**.

     [MP] From **A** and **A ⊃ B**, infer **B**.

     [UG] From **B ⊃ A(y/x)**, infer **B ⊃ [x]A**, provided **y ∈ (y/x)** and y is not free in the latter sentence.

      [RI]   **[x]x = x** is an axiom.

     [LL]   **[x][y](x = y ⊃ (A ⊃ A(y/x)))** is an axiom if **y ∈ (y/x)**.

These postulates along with **[E]** constitute **free logic (FL)** for the language **L**. The notions of proof and theorem are as usual, and ↦ **A** abbreviates "**A** is a theorem." A **theory** in **FL** is any set of statements. If **T** is a theory, **A** is **deducible in**, or **a theorem of**, **T** (**T** ↦ **A**) if and only if for some **B₁...**, **Bₙ ∈ T, B₁ ⊃ (Bₙ ⊃ A)...)** is a theorem of **FL**. **T** is **consistent** if and only if not every statement is a theorem of **T**.

    The soundness of **FL**, in the sense that every theorem is universally valid, is readily proved by induction on length of proof using **Lemma 3** and **Lemma 4**. Hence every **satisfiable** theory is consistent; the rest of this section establishes the converse.

    Assume as known the proof that every consistent theory **T** of **L** can be extended to a theory **T*** of a **saturated** extension **L*** of **L** in the sense that (a) **T*** is consistent but no proper extension of **T*** is consistent and (b) for every statement **[x]A** of **L***, **[x]A ∈ T*** if and only if for every constant **c** of **Lₙ**, **A(c/x) ∈ T***. Assume also the usual construction of a canonical standard model **<D*,v*>** for classical identity theory in **L*** in which **D*** is the set of equivalence classes |t| of closed terms under the relation **t = s ∈ T***. It follows from saturation that if **[∃x]A ∈ T***, then **A(c/x) ∈ T*** for some constant **c**; in view of the theorem **[∃x]x = t**, this implies that **c = t ∈ T*** and hence |t| = |c| for some **c**, for every

---

via reduction to the classical logic of function symbols. Here a similar proof is possible, but instead the discussion proceeds more directly via standard Henkin techniques.

closed term $\mathbf{t}$. So, now and in the sequel, let $\mathbf{D}^*$ be the set of $|\mathbf{c}|$ for $\mathbf{c}$, a constant of $\mathbf{L_N}$. Assume in fact that the displayed constant is the alphabetically first constant in $|\mathbf{c}|$ and thus uniquely determined.

$<\mathbf{D}^*, \mathbf{v}^*>$ is extended to a free description model $M^* = <\mathbf{D_1}^*, \mathbf{D_2}^*, I^*, \mathbf{EX}^*>$ by setting

$\mathbf{D_2}^* = \mathbf{D}^*$,

$\mathbf{D_1}^* = \{|\mathbf{lc}|: \mathbf{E!}(\mathbf{c}) \in \mathbf{T}^*\}$,

$\mathbf{EX}^* = \{(\mathbf{skel}(\mathbf{x}, \mathbf{A}), \mathbf{g}): (\mathbf{x}, \mathbf{A})$ is a @complex predicate of $\mathbf{L}$ and $\mathbf{g}$ is a function from $\mathbf{D}^*$ to subsets of $\mathbf{D}^*$ of the same degree as $\mathbf{skel}(\mathbf{x}, \mathbf{A})\}$

$I^* = \lambda(\mathbf{skel}(\mathbf{x}, \mathbf{A}), \mathbf{g})(\lambda|\mathbf{c_1}|...|\mathbf{c_k}|(l(\mathbf{skel}(\mathbf{ix})\mathbf{A})(\mathbf{c_1}...\mathbf{c_k}/\mathbf{n_1}...\mathbf{n_k})l))$

Define a function $\mathbf{f}$ on the closed expressions of $\mathbf{L}^*$ as follows: If $\mathbf{E}$ is a statement, $\mathbf{f(E)} = \mathbf{t}$ if $\mathbf{E} \in \mathbf{T}^*$, otherwise $\mathbf{f(E)} = \mathbf{f}$; if $\mathbf{E}$ is a singular term, $\mathbf{f(E)} = |\mathbf{E}|$; if $\mathbf{E}$ is a @complex predicate, say $(\mathbf{x}, \mathbf{A})$, $\mathbf{f(E)} = \{|\mathbf{lc}|:\mathbf{A(c/x)} \in \mathbf{T}^*)$. Finally, if $\mathbf{E}$ is a k-ary skeleton, let $\mathbf{f(E)} = \lambda|\mathbf{c_1}|...|\mathbf{c_k}|$ $\mathbf{f(E(c_1...c_k/n_1...n_k)})$. (That this is well-defined follows from [LL] and saturation, as does the well-definedness of I^* above.)

Now for $\mathbf{x_1...x_n}$, the free variables of $\mathbf{E}$, and an assignment α such that $\alpha\mathbf{x_i} = |\mathbf{c_i}|$, it will be stipulated that $\mathbf{v}^*\alpha(\mathbf{E}) = \mathbf{f(E(c_1...c_n/x_1...x_n)})$. Then every $\mathbf{v}^*\alpha$ satisfies conditions [1]–[I0]. The proof here is restricted to the cases $\mathbf{E} = (\mathbf{y})\mathbf{A}, \mathbf{E} = (\mathbf{y}, \mathbf{A}))$, $\mathbf{E}$ is a skeleton, and $\mathbf{E} = (\mathbf{iy})\mathbf{A}$, since the other cases are familiar from classical predicate logic. For simplicity, abbreviate the substitution in the definition in the first sentence of this paragraph by $(\mathbf{c/x})$.

In the first case, abbreviating "if and only if" as "iff",

$\mathbf{v}^*(\mathbf{y})\mathbf{A} = \mathbf{t}$ iff $(\mathbf{y})\mathbf{A(c/x)} \in \mathbf{T}^*$

         iff $[\mathbf{y}](\mathbf{E!}(\mathbf{y}) \supset \mathbf{A(c/x)}) \in \mathbf{T}^*$ (by [E] and saturation)

         iff $\mathbf{E!}(\mathbf{d}) \supset \mathbf{A(cd/xy)} \in \mathbf{T}^*$ for every constant $\mathbf{d}$ (by saturation)

         iff for all $|\mathbf{d}| \in \mathbf{D_1}$, $\mathbf{f(A(cd/xy))} = \mathbf{t}$, (by definition of $\mathbf{D_1}$ and saturation)

         iff for all $|\mathbf{d}| \in \mathbf{D_1}$, $\mathbf{v}^*\alpha(\mathbf{ldl/y})\mathbf{A} = \mathbf{t}$.

In the second case,

$\mathbf{v}^*(\mathbf{y,A}) = \mathbf{f(y,A(c/x))}$

      $= \{|\mathbf{d}|: \mathbf{A(c/x)(d/y)} \in \mathbf{T}^*\}$ (by definition of $\mathbf{f}$)

$$= \quad \{|\mathbf{d}|: \mathbf{A}(\mathbf{cd/xy}) \in \mathbf{T}^*) \text{ (by properties of substitution)}$$
$$= \quad \{|\mathbf{d}|: \mathbf{v}^*\alpha(\mathbf{ldl/y})\mathbf{A} = \mathbf{t}\} \text{ (by definition of } \mathbf{v}^*).$$

In the third case,

$$\mathbf{v}^*\alpha\mathbf{E} = \quad \lambda|\mathbf{d}_1|...\mathbf{ld}_n|\mathbf{f}(\mathbf{E}(\mathbf{d}_1...\mathbf{d}_k/\mathbf{n}_1...\mathbf{n}_k))$$
$$= \quad \lambda|\mathbf{d}_1|...|\mathbf{d}_n|\mathbf{f}(\mathbf{E}(\mathbf{y}_1...\mathbf{y}_k/\mathbf{n}_1...\mathbf{n}_k)(\mathbf{d}_1...\mathbf{d}_k/\mathbf{y}_1...\mathbf{y}_k)) \text{ (by prop-erties of substitution, where the } \mathbf{y}_i \text{ are the first } \mathbf{k} \text{ variables not in } \mathbf{E})$$
$$= \quad \lambda|\mathbf{d}|\mathbf{v}^*\alpha(\mathbf{ldl/y})\mathbf{E}(\mathbf{y/n}) \text{ (abbreviating, and using definition of } \mathbf{v}^*, \text{ since the } \mathbf{y}_I \text{ are all the free variables of } \mathbf{E}(\mathbf{y/n})).$$

In the last case,

$$\mathbf{v}^*\alpha(\mathbf{iy})\mathbf{A} = \quad \mathbf{l(iy)A(c/x)l}$$
$$= \quad \mathbf{l(skel(iy)A(c/x))(tk...tk/n}_1...\mathbf{n}_k)| \text{ (by the definition of skeleton, where } \mathbf{t}_1...\mathbf{t}_k \text{ are the arguments of } (\mathbf{iy})\mathbf{A(c/x)}),$$
$$= \quad \mathbf{l(skel(iy)A)(s}_1(\mathbf{c/x})...\mathbf{s}_k(\mathbf{c/x})/\mathbf{n}_1...\mathbf{n}_k) \text{ (by \textbf{Lemma 1}, where } \mathbf{s}_1...\mathbf{s}_k \text{ are the arguments of } (\mathbf{iy})\mathbf{A})$$
$$= \quad \mathbf{l(\underline{skel}(iy)A)(d}_1...\mathbf{d}_k/\mathbf{n}_1...\mathbf{n}_k) \text{ (by saturation and [\textbf{LL}], where } |\mathbf{s}_i(\mathbf{c/x})\mathbf{l}...|\mathbf{d}_i|$$
$$= \quad (\lambda(|\mathbf{d}_1|...|\mathbf{d}_k|(|\mathbf{skel(iy)A})(\mathbf{d}_1...\mathbf{d}_k/\mathbf{n}...\mathbf{n}_k)|(|\mathbf{s}_1(\mathbf{c/x})|...|\mathbf{s}_k(\mathbf{c/x})\mathbf{l}) \text{ (by } \lambda\text{-\textbf{abstraction})}$$
$$= \quad (I^*(\mathbf{skel(y,A)}, \mathbf{v}^*\alpha(\mathbf{skel(y},(|\mathbf{s}_1(\mathbf{c/x})|...|\mathbf{s}_k(\mathbf{c/x})|)(\text{by def-inition of } I^*, \text{ since by the definitions of } \mathbf{v}^* \text{ and } \mathbf{EX}^*, (\mathbf{skel(y, A)}, \mathbf{v}^*(\mathbf{skel(y, A)})) \text{ is in } \mathbf{EX}^*)$$
$$= \quad (I^*(\mathbf{skel(y, A)}, \mathbf{v}^*\alpha(\mathbf{skel(y, A)})))(\mathbf{v}^*\alpha\mathbf{s}_1...\mathbf{v}^*\alpha\mathbf{s}_k).$$

It follows immediately that $\mathbf{M}^*$ satisfies all of $\mathbf{T}^*$ and hence also $\mathbf{T}$. This finishes the proof sketch that every consistent theory $\mathbf{T}$ is satisfiable.

7. A MINIMAL DEFINITE DESCRIPTION THEORY AND UNIFORM COMPLETENESS PROOFS

As indicated in the Introduction, all definite description theories are based on the principle

$$\mathbf{P}_0 \quad (\exists_1\mathbf{x})\mathbf{A} \supset \mathbf{A}((\mathbf{ix})\mathbf{A/x}),$$

a generalization of (1). Let $\mathbf{MD}$ be the result of adding all instances of $\mathbf{P}_0$ as axioms for which $\mathbf{A} \in (\mathbf{ix})\mathbf{A/x}$. This final section proves the completeness of **MD** relative to the following constraint on models:

C_1 If $<E, g> \in EX$, and $f(g(d_1...d_k)) = \{d\}$,
then $(I(E, g))(d_1...d_k) = d$.

Given C_1, essentially a restriction on the *specification function I*, the following proof evidently illustrates a uniform method by which all of the systems in the appendix may be proved (strongly) complete.

Suppose that **T** is a consistent theory in **MD**. Then **T** with the universal closures of the P_0 axioms is a theory **TD** of **FL** and is consistent in **FL**. Hence by the reasoning in the previous section, there is a model $M_{TD} = <D1, D2, I, EX, v>$ of TD. Let **EX'** be the set of hyper-intensions $<skel(x, A), v(skel(x, A))>$, **A** be a sentence of **L**, and let I' be restricted to **EX'**. Then $M'_{TD} = <D1, D2, I', EX', v>$ and determines the same extended valuation as M_{TD}. So it may be presumed in M_{TD} that every member of **EX** has the indicated form.

Hence if $<E, g> \in EX$, then for some **A** and **x**, $E = skel(x, A)$ and $g = v(skel(x, A)) = \lambda(d_1...d_k(v\alpha(d_1...d_k/y_1...y_k)skel(x, A)(y_1...y_k/n_1...n_k))$, for any α the y_i being as in [9]. If $g(d) = \{d'\}$ for some **d'**, and $skel(x, A)(y/n) = (x, B)$, then $v\alpha(d/y)(e/x)B = T$ if and only if $e = d'$. The y_i are all the free variables of **(ix)B**. Since none of them occur in $skel(x, A)$, none of them are bound in **B**, and, therefore, $B \in (ix)B/x$. Hence, since **M** satisfies P_0, $v\alpha(d/y)(B((iy)B/x) = T$. So by semantic substitution $v\alpha(d/y)((v\alpha(d/y)(ixB/x)B = T$, and, hence by the preceding, $v\alpha(d/y) (ix)B = d'$. But by [10], $v\alpha(d/y) = (I(skel(x, B), v(skel(x, B)))(v\alpha(d/y)t)$ where t_1, etc. are the arguments of **(x, B)**. Again, by construction, the arguments of **(x, B)** are $y_1...y_k$. Hence $v\alpha(d/y)t = d$, and by **Lemma 1**, $skel(x, B) = skel(x, A)$; so $(I(E,g))(d) = d'$, as required. Therfore **M** satisfies the constraint C_1, and is a model for **T**.

The method just used is general. It would appear that one can construct theories of positive free definite descriptions *a la mode* (see the Appendix) by selecting axioms to be added to those of **MD**, and impose corresponding constraints on models with the assurance that the resulting systems will be sound and complete.

8. APPENDIX

What follows is a list of the various axiom schemata and/or rules, which, when added to **MD**, yield various positive free description theories. Most of them have appeared in the literature, but some are presented here for the first time. The list is split into two kinds of axioms, the **P-class** (having to do with the conditions under which a predicate **F** may be asserted of the entity purported to be described by **(ix)F(x)**, and the **I-class** (having to do with the conditions under which identities including definite descriptions with non-unique bases can be asserted). That two kinds of axioms are (and ought to be) distinguished, reflects the fact that the hierarchy is essentially two dimensional, and is crucial to the observation made earlier in this chapter that van Fraassen's picture of the hierarchy as a linearly ordered sequence of (proof theoretical) inclusion relations is incorrect. Indeed, the system of Grandy (see below) cannot be included van Fraassen's original one dimensional hierarchy.

The **P** axiom schemata and rules are the following

P₀ $(\exists_1 x)A \supset A((ix)A/x),$

P₁ $E!(ix)A \equiv (\exists_1 x)A$ [25]

P₂ $t = (ix)(x = t)$ [26]

P₃ $\dfrac{A \supset x = t}{A(t/x) \supset A((ix)A/x)}$ if **x** is not free in **t**

P₄ $\dfrac{B \supset (A \supset x = t)}{B \supset ((A(t/x) \supset A((ix)A/x))}$ if **x** is not free in **B** or **t** [27]

P₅ $(A(t/x) \ \& \ \sim E!(t)) \supset A((ix)A/x)$

P₆ $(A(t/x) \ \& \sim (\exists x)A) \supset A((ix)A/x)$

P₇ $(A(t/x) \ \& \ (x)(A \supset x = t\,)) \supset A((ix)A/x)$

The **I** axiom schemata and rules are the following.

I₀ $s = t$ if **s** is an alphabetic variant of **t**

I₁ $(x)(A \equiv B) \supset (ix)A = (ix)B$

---

[25] In most if not all positive free definite description theories **P₀** and **P₁** are jointly equivalent to $(x)(x = (iy)A \equiv (y)(A \equiv y = x))$. Most presentations of free definite description theory, traditionally, have taken this latter formula as expressing the fundamental property of definite descriptions, It has come to be known in the free logic literature as *Lambert's Law.*

[26] Notice that this is virtually a special instance of **A((ix)A/x)**.

[27] **P₄** and $(A(t/x) \ \& \ [x](A \supset x = t)) \supset (A(ix)A/x)$ are mutually deducible in **FL**.

I_2 $C \supset (A \equiv B)$

———————— if **x** is not free in **C**

$C \supset (ix)A = (ix)B^{28}$

I_3 $[y_1]...[y_n](x)(A \equiv B) \supset [y_1]...[y_n] (ix)A = (ix)B$

I_4 $\sim(\exists_1 x)A \supset (ix)A = (ix)(x \neq x)$

The various **systems** of positive free definite description theory are obtained by adding various **P** and/or **I** axiom schemata and rules to **FL**. Van Fraassen's initial one dimensional hierarchy consisted of

MFD $(= \mathbf{FL} + \mathbf{P_0} + \mathbf{P_1})$ **Lambert and van Fraassen**
FD1 $(= \mathbf{MFD} + \mathbf{P_2})$ **Lambert**
FDL $(= \mathbf{MFD} + \mathbf{P_7})$ **Lambert**
FDV $(= \mathbf{MFD} + \mathbf{I_1})$ **van Fraassen**
FD2 $(= \mathbf{MFD} + \mathbf{FL_1}(= \mathbf{FL} + (\sim\mathbf{E!}(s) \& \sim\mathbf{E!}(t) \supset s = t))$ **Lambert**

Some systems, e.g.,

FDS $(= \mathbf{MFD} + \mathbf{I_4})$ **Scott**

and

FDT $(= \mathbf{MFD} + \mathbf{P_5})$ **Thomason,**

can be sandwiched into van Fraassen's one-dimensional hierarchy, but the system

FDG $(=\mathbf{MFD} + \mathbf{I_2})$ **Grandy**

cannot. Though a theorem in the strongest theory **FD2** it is not a theorem in **FDV** (but **FDV** is also a theorem in **FD2**). The current picture of the hierarchy is two-dimensional with **MFD** constituting the minimal theory and **FD2** the maximal theory on either the **P**-dimension or the **I**-dimension.

The theories represented by the addition to **MFD**, respectively, of **P_3, P_5, I_0,** and **I_3** are inventions of **Lambert and Woodruff**. The last mentioned theory may be unique in requiring an **[x]** ("outer") quantifier.

_____

[28] I_2 and $[x](A \equiv B) \supset (ix)A = (ix)B$ are mutually deducible in **FL**.

6

Predication and Extensionality

1. INTRODUCTION

Quine's concept of predication is intimately related to his notion of referential opacity. To treat the position of a singular term in a sentence as "purely referential", and hence the sentence as "referentially transparent", is to treat the very same sentence as a *predication*.[1] "Predication", he says, "joins a general term and a singular term to form a sentence that is true or false according as the general term is true or false of the object, if any, to which the singular term refers."[2] The conception of predication expressed in the quoted passage is not restricted to its author. Among others who hold it are many free logicians.

Quine has also written that "so long merely as the predicated general term is true of the object named by the singular term ... the substitution of a new singular term that names the same object leaves the predication true."[3] So if a sentence is a predication, it satisfies the *substitutivity of identity*. Moreover, "in an opaque construction you ... cannot in general supplant a general term by a *coextensive* term (one true of the same objects) ... without disturbing the truth-value of the containing sentence. Such a failure is one of the failures of extensionality."[4]

[1] W. V. Quine, *Word and Object*, Wiley, New York (1960), p. 96.
[2] Ibid., p. 96.
[3] Ibid., p. 151.
[4] Ibid., p. 142.

92

The theory of predication under consideration is *non-extensional* precisely in the sense that it does not satisfy the principle that *co-extensive general terms substitute for each other salva veritate.*[5] The proof of this fact is the first order of business. On the other hand, if it did satisfy this principle, a principle one would expect a confirmed extensionalist such as Quine to demand, then that theory is easily proved unsound. So the second objective is to elicit some of the implications of the claim that the theory of predication under discussion is non-extensional. The final objective is to show how the putatively unsound version of the theory might be fixed within certain constraints that Quine himself does, or presumably would, acknowledge.

2. PREDICATION AND EXTENSIONALITY

The sentences of immediate concern are in nearly, but not yet regimented language, natural or otherwise. The reason for this fussy sounding description is that it is controversial that natural language contains the free variables **x, y, z, . . .** But *before* regimentation Quine at least treats both the sentences

x broods

and

Nixon broods

as predications. Though this section is concerned mainly with the *general* theory of predication Quine's words are used to express, a Quinelike reaction to the non-extensionality of that theory will be discussed later. Hence the need in the present discussion to treat sentences in at least one free variable.

---

[5] Another way of characterizing non-extensionality has been adopted by Nuel Belnap in his foreword to Aldo Bressan's book *A General Interpreted Modal Calculus,* Yale University Press, New Haven (1973). In Belnap's terminology the initial argument here is intended to show that even in a non-modal, non-epistemic language in which (1) at least one singular term **t** is not assigned an element of the domain of discourse (construed as the set of existents), (2) the predicates take sets of n-tuples from the domain of discourse as extensions, (3) there is a one place predicate **P** such that **P(t)** is false, the truth-value of a statement containing **t** does not depends on just the extensions of its constituent terms. (See the next chapter.)

How about a sentence such as

Vulcan rotates (on its axis)?

Is this sentence a predication? The answer is yes, and that verdict is crucial to the ensuing argument. So it won't hurt to see why this sentence counts as a predication, especially since Quine's way of expressing the present theory of predication is not transparent on this score.

Consider, again, the second of the listed sentences above containing as its subject term one that has existential import. How does one tell that it is a predication? Because it satisfies the *substitutivity of identity*.[6] Consider next the third sentence listed above containing as its subject term one that does not have existential import. It is a predication because it also satisfies the *substitutivity of identity*. This is so even though its falsehood cannot depend on the general term 'rotates' being false of Vulcan, there being no such entity. Russell would have said, "Exactly!" – no pun intended – "and that is *why* it is not a predication." But this is not Quine's view, nor is it the view of anyone else who espouses the theory of predication lately expressed in Quinely prose.[7]

---

[6] This is not so in the case of the sentence 'The commissioner is looking for the chairman of the hospital board', opaquely interpreted. Suppose it unknown to the commissioner that the dean is the chairman of the hospital board. Then even though (i) the expression 'the chairman of the hospital board' is a singular term, (ii) what is left of the sentence, namely, the expression 'The commissioner is looking for', is a general term, and (iii) the resulting sentence is true, the sentence is not a predication. Why? Because the general term 'The commissioner is looking for' is not true of the *object* specified by the singular term 'the chairman of the hospital board'. If it were, then it would also be true of the same object no matter how specified, and in particular specified by the co-referential singular term 'the dean'. And this contradicts the assumption that 'looking for' is opaquely interpreted.

[7] A potentially troublesome feature arising in the classification of sentence like 'Vulcan (is an object such that it) rotates' as a predication has to do with the meaning of 'co-referential'. The meaning of this word is perfectly clear when singular terms having existential import are the topic of concern; if the singular terms **a** and **b** have existential import, they are co-referential just in case **a** refers to the same thing to which **b** refers. But when singular terms not having existential import (hence which for Quine don't refer) are the topic of concern, the meaning of 'co-referential' is ambiguous, and the status of sentences such as 'Vulcan rotates' as predications may not seem so obvious. To say that **a** and **b** are co-referential (refer to the same thing) when they do not have existential import can mean that **a** and **b** refer *and* **a** refers to whatever is the referent of **b**. Or it can mean, more weakly, just that **a** refers to whatever is the referent of **b**. (In the case of referential singular terms these meanings collapse.)

For others who support the view that 'Vulcan rotates' is a predication, the reason it so qualifies must be something like this. Logical form, in part at least, concerns the logical roles played by singular and general terms in sentences, and the expression 'Vulcan' in the sentence 'Vulcan rotates' certainly plays the same logical role as the expression 'Nixon' in the sentence 'Nixon broods'. So the sentence 'Vulcan rotates' qualifies as a predication because the truth or falsity of that sentence *would* depend on whether the general term 'rotates' is true (or false) of the referent of the singular term 'Vulcan' *were* the word 'Vulcan' to refer. I don't for a minute say that Quine would ever have subscribed to this last statement of motivation; but I would be surprised if many free logicians found it unacceptable. At any rate, these considerations point up the importance of the rider expressed in the words 'if any' in Quine's definition of predication; they grant a special dispensation to sentences such as 'Vulcan rotates'.

Another important preliminary point is that predications may be simple or complex. In Quine's case at least there is straightforward acknowledgment of this fact on page 175 of his book *Word and Object*. So it would be wrong to think of the predications merely as the atomic sentences of the language. The sentences 'Nixon broods' and 'Nixon is a secretive man' both are predications; and so is the sentence 'There is a planet that is Vulcan'.

In addition to the assumptions that

(A1) the sentence 'Vulcan rotates' is a genuine predication

and

(A2) predications can be complex,

On the stronger understanding if **a** is co-referential to **b**, the sentence 'a = b' is false even if **a** and **b** are the same irreferential singular term. On the weaker understanding all such identities are true, and might lead one to believe that 'Vulcan flies' is not a predication. For suppose the sentence in question is counted truth-valueless, as Quine is disposed to count it. Then the intuitions of some compel a false verdict in the case of the 'The wingless gold mountain flies' even though, on the weaker understanding of 'co-referential', 'Vulcan is the wingless gold mountain' is true. Quine's own preference is nowhere explicitly expressed in *Word and Object* as far as I know, but his insistence on the predicational status of a sentence such as 'Vulcan rotates' and the attitude displayed in the chapter entitled 'Regimentation' compel the stronger interpretation.

the argument that the theory of predication under examination is non-extensional depends on the following pair of assumptions:

> **(A3)** There is a general term true of every object but which yields a false sentence when prefixed by a singular term without existential import,

and

> **(A4)** Any sentence satisfying **(A3)** is a predication.

For Quinians, the sentence 'Vulcan exists' satisfies both **(A3)** and **(A4)**. It will be used in what follows to help fix the ideas. But supporters of the present theory of predication who object to the word 'exists' as a predicate, and thus to the sentence 'Vulcan exists' as a predication may substitute the expression 'identical with something' for the word 'exists', a general term that is a predicate on pain of incoherence. For if complex general terms containing identities are not predicates it is difficult to understand what general terms would be. The reason for belaboring this issue about the word 'exists' is that nothing in the argument that the present theory of predication is non-extensional turns on the word 'exists' being taken as a predicate, logical or otherwise; hence the statement of **(A3)** in the more general if less comprehensible fashion.

Consider again the sentence 'Vulcan rotates' and assume that it is a predication. Consider also the general terms 'rotates', 'existent that rotates' and 'rotates if an existent'. Assume that the sentences 'Vulcan rotates' and 'Vulcan exists' are genuine predications, and that complexes made up of predicates by the usual logical operations of conjunction and conditionalization are also predicates, which surely is the case for Quine at least. Then the three general terms mentioned a sentence back are predicates. They are also *co-extensive* because each of these predicates is true (or false) of the same objects, namely, everything that rotates (or nothing that rotates). What shall now be established is the following: "No matter what truth-value the sentence 'Vulcan rotates' has, if any, substitution of a coextensive predicate for the predicate 'rotates' in the sentence 'Vulcan rotates' does not always preserve truth-value."

There are three cases to consider.

Case 1: 'Vulcan rotates' is *true*. Then substitution of the general term 'existent that rotates' for the general term 'rotates' in the sentence

'Vulcan rotates' yields the false sentence 'Vulcan is an existent that rotates'.

Case 2: The sentence 'Vulcan rotates' is *false*. Then substitution of the general term 'rotates if an existent' for the general term 'rotates' in the sentence 'Vulcan rotates' yields the *true* sentence 'Vulcan rotates if an existent'.

Case 3: The sentence 'Vulcan rotates' is *truth-valueless*. Then substitution, for example, of the general term 'existent that rotates' for the general term 'rotates' in the sentence 'Vulcan rotates' yields the *false* sentence 'Vulcan is an existent that rotates'.

It follows that the present theory of predication is non-extensional as Quine has characterized that term in his book *Word and Object*.[8]

3. SOME IMPLICATIONS

The argument presented in the previous section does not require the particular predicate 'rotates'; any predicate will do. It will go through even in a non-modal, non-epistemic language containing identity as its only atomic predicate. Just replace the general term 'rotates' by the general term 'self-identical', and the general term 'exists' by the general term 'is identical with something'.

The argument that the present theory of predication is non-extensional is simply an application of the reasoning to show that the classically valid principle:

(1) $\forall x(F(x) \equiv G(x)) \supset (F(t/x) \equiv G(t/x))$, where t is a singular term,

fails in free logic. This principle is deductively equivalent to the classical principle of *Specification*,

(2) $\forall x(F(x)) \supset F(t/x)$, where t is a singular term,

a principle that holds in any free logic only on the condition that t is a singular term with existential import.

The failure of (1) is illuminating when comparing singular terms such as 'Vulcan' with what Thomason calls "substance terms"; the former cause a breakdown in the principle of the *substitutivity of coextensive*

---

[8] Op cit., *Word and Object*, p. 151.

predicates and non-substance terms cause a breakdown in the *substitutivity of identity.*[9]

Another application of the present argument has to do with logical form. In his essay, 'What is Russell's theory of descriptions', David Kaplan presents a sketch of his view of logical form and characterizes in a rough and ready way the concept of a logically perfect language.[10] Kaplan says that "logical form is determined by the *evaluation rules* of the language. These rules tell us how to 'construct' the semantical value of an expression in terms of the values of its logically simple components". He writes,

'For logical perfection we require that the logically simple expressions coincide with the grammatically simple (but well-formed) expressions, and that to every formation rule there corresponds a unique evaluation rule, such that any compound formed by applying the formation rule to given components is evaluated by the corresponding evaluation rule to the values of the components'; in short, in a logically perfect language 'the semantical evaluation rule of an expression recapitulates its grammatical construction'.[11]

Russell's theory of descriptions is not logically perfect in Kaplan's sense because there is no corresponding semantical evaluation rule

[9] R. H. Thomason, 'Modal Logic and metaphysics' in *The Logical Way of Doing Things* (ed. Karel Lambert), Yale University Press, New Haven, CT (1969). It is also interesting to note that the principle of uniformity helping to dictate Bressan's treatment of singular terms in modal logic has its counterpart in some developments of free logic. Just as variables in Bressan's development are denied the privileged treatment of substance terms, so, in some developments of free logic, the principle of uniformity requires that variables be disallowed the privileged status of existential import. And this decision *can* have important consequences. For example, in his paper, 'A Definition of Truth for Theories with Intensional Description Operators', *Journal of Philosophical Logic, 1* (1973), p. 137, Richard Grandy, in an observation inspired by David Kaplan, points out that unless one adopts the policy that variables do not always have existential import, the stronger looking rule

$$\frac{F(x) \equiv G(x)}{iy(F(y)) = iy(G(y)),}$$

in free definite description theory turns out to be equivalent to the clearly extensional rule

$$\frac{\forall x(F(x) \equiv G(x))}{iy(F(y)) = iy(G(y)).}$$

[10] Kaplan's essay appears in *Physics, History and Logic* (eds. W. Yourgrau and A. Breck), Plenum Press, New York (1970), pp. 277–288.
[11] Ibid., p. 283.

for determining the value of sentences containing those syntactically well-formed expressions called definite descriptions. But, notes Kaplan, this does not prevent the construction of logically perfect languages containing definite descriptions as terms satisfying the truth conditions reflected in Russell's famous contextual definitions of descriptions in *14 of *Principia*. He says there are at least two ways of accomplishing this objective depending upon whether improper (unfulfilled) definite descriptions are assigned nothing or something. In the first case, one may simply rewrite the classical evaluations for atomic sentences containing terms, that is, variables, individual constants, *or* definite descriptions. In the simplest case this is as follows:

Where **F** is a predicate and **t** is a singular term, (i) **Ft** is true just in case **t** denotes some-thing that is **F** and (ii) **Ft** is false just in case it is not true.

In the second case, the case in which improper definite descriptions are assigned something but *outside* the domain of discourse, logical perfection is achieved for atomic sentences as follows:

Where **F** is a predicate and **t** is a singular term (i) **Ft** is true just in case **t** denotes something that is **F** and is in the domain of discourse and (ii) **Ft** is false just in case **t** denotes something that is not **F**.

Moreover, even free definite description theories that count some atomic sentences containing improper descriptions true (contra Russell) are logically perfect *à la* Kaplan, for example, the free definite description theory in Dana Scott's paper 'Existence and description in formal logic'.[12]

An interesting consequence of Kaplan's theory of logical perfection is that it does not entail that a logically perfect language is extensional – in Quine's sense of the word. For the case where improper descriptions are assigned *nothing* in a logically perfect language, introduce a monadic predicate **P** such that its extension is the domain of discourse itself. Then, by the evaluation rules for atomic sentences in such a language as pictured by Kaplan, the sentence

Pix(Fx & ~Fx), where **i** is the definite description operator,

---

[12] See his paper in *Bertrand Russell: Philosopher of the Century* (ed. R. Schoenman), London (1967).

is false, assuming that the definite description in question is assigned nothing. Given the classical interpretation of the atomic predicates, and assuming the classical evaluation rules for compound statements, it is clear that an arbitrary atomic predicate **G** is coextensive with the complex predicate **G if P**. So Case 2 in the argument above applies. Hence, logical perfection does not imply extensionality even when the evaluation rules in question deal with extensions only in the sense of assigning n-tuples of individuals to n-adic predicates, individuals to names, and truth-values to sentences.

What is the effect of the major non-extensionality argument in this essay on Quine's own theory of singular terms? In sections 37 to 39 of his book *Word and Object*, Quine restates and defends his influential theory on the propriety of eliminating singular terms. He notes, in section 37, that though the position of 'Pegasus' in 'Pegasus flies' satisfies the *substitutivity of identity*, nevertheless it is anomalous (though not contradictory) to admit that 'Pegasus' has purely referential position therein. He also notes that alleged truth-value gaps enter with non-referential singular terms, and so on. The clear suggestion is that 'Pegasus' – the expression – is an inconvenience, an embarrassment to good philosophical taste which favors the austere and transparent over the baroque and opaque, and, therefore, should be banished from "the canonical language." Sentences in colloquial discourse containing singular terms without existential import are to be paraphrased via the Russell-Quine elimination method into the canonical language.

To paraphrase Russell, it is not customary for a philosopher to face 'Pegasus' with so much courage, and indeed not even the canonical language can withstand its onset. The argument is that retaining expressions such as 'Pegasus' (or 'Vulcan') is more than inconvenient; 'Pegasus', the expression, is the Devil himself, and, in the words of the Scriptures, *must* be bound in chains and cast down. For one can expect Quine to hold that the position of general terms in predications obeys the principle of the *substitutivity of coextensive general terms*, an expectation which, in fact, he has acknowledged is correct.[13] So, singular terms without existential import must of necessity be disallowed occupancy in that heavenly language whose purity is transparent; and vice versa.

---

[13] In a personal letter to me in April 1973.

In section 39 of his book *Word and Object*, a section prophetically entitled 'Definition and the Double Life', Quine observes that the unavailability of constant singular terms in the canonical language has nagging disadvantages, for example, the hamstringing of mathematics because of inability to "bandy names as singular terms, and descriptions likewise, substituting them for variables and predicatively applying general terms to them" sacrificing "precisely the moves that typify mathematics at its fleetest."[14] "Happily," he says, "the looming dilemma," the problem of truth-value gaps, an anomalous concept of pure referentiality (and most importantly, as of this essay, the non-extensionality of predication), versus the hamstringing of mathematics, "can be solved."[15] A shorthand use of the previously excised singular terms can be *defined* relative to the canonical notation eschewing such singular terms. For example, a rehabilitated 'Vulcan' can be reintroduced as an abbreviation of 'the thing that Vulcanizes'. Then contexts containing the rehabilitated singular term, for example, the sentence 'Vulcan rotates', can be rendered in turn via Russell's method as an abbreviation of the sentence 'There is one thing such that exactly it vulcanizes and flies'.

This procedure for reintroducing singular terms other than variables into the canonical language does abolish truth-value gaps; atomic sentences containing singular terms without existential import, for example, all turn out false. But what about the logical status of the rehabilitated sentences 'Nixon Broods' and 'Vulcan rotates'? Are these to count as predications, though reformed ones? This question poses a new dilemma just as upsetting as the dilemma Quine's resuscitation of singular terms via definitions is intended to avoid.

Suppose such sentences are counted as predications. Then, in virtue of the predication 'Vulcan rotates', an anomalous concept of pure referentiality is resurrected, and the non-extensionality of the theory of predication returns to haunt just as forcefully as before. On the other hand, suppose such sentences are not regarded as predications. Then the sense in which the expressions 'Nixon', 'Vulcan' (or 'Pegasus') are singular terms is a sense not appropriate for mathematics on

---

[14] Ibid., p. 188.
[15] Ibid., p. 188.

Quine's own account because it is a sense in which general terms, by agreement, are not applicable predicatively to them. Such philosophical double talk, to paraphrase a prestigious twentieth century philosopher, repudiating singular terms while enjoying their benefits, thrives no doubt on the looseness of the term 'definition'.[16]

4. FIXING THE THEORY

Consider the sentence

(3)   Vulcan is an object **x** such that **x** rotates.

This sentence is the result of applying the logical rule of *predicate abstraction*, a rule that regiments the move from the sentence 'Vulcan rotates' to another with the "such that" construction.[17] It specifies what is being predicated of Vulcan (or of the purported referent of any singular term in subject position), namely, the predicate

   object **x** such that it **x** rotates,

or if converted to verb form,

   is such that object **x** is such that **x** rotates

(3) satisfies the *substitutivity of identity* and, hence, is a predication. But, given Quine's muscular extensionalism, it is at least a necessary condition of a sentence's being a predication that it violate no extensionality principle. On the other hand, (3), and its stylistic variants are not predications because they violate the *salva veritate substitution of co-extensive predicates*. It follows that Quine's theory is unsound. Of course, if the theory is revised to require satisfaction of *all* extensionality conditions, then the sentence in question will simply fail to be a predication at all, a verdict that seems to conflict with the impulse behind Quine's definition of 'predication'.

The cut-the-losses alternative to the apparent unsoundness in Quine's theory is to eliminate all sentences containing singular terms without existential import from the category of predications. This amounts to removing the condition "if any" in the earlier statement

---

[16] Ibid., p. 242.

[17] W. V. Quine, *Methods of Logic* (fourth edition), Harvard University Press, Cambridge, MA (1982), pp. 133–134.

of Quine's theory and giving up assumption (**A1**). The resulting class of sentences is completely extensional because it is a subclass of the class of instances in standard first order predicate logic, a class easily proven to be completely extensional in the above sense.

This course of action has immediately gratifying benefits. The position(s) of the singular terms in a predication are no longer anomalous because all such positions get filled only by referential singular terms. Nor is it inconvenient because the sentences that might occasion a multi-valued (or perhaps a two valued semantical underpinning with truth-value gaps) – (3), for instance – no longer arise. Of course the puzzlement Quine associates with predications such as

(4)   The attribute of being greater than 4 is necessary of 9

still hovers because the singular terms in (4) all have existential import.[18] But the problem here has less to do with Quine's notion of predication than with the modal phrase 'is necessary of'. In a language without phrases like this, or phrases like 'is believed to be', and so on, it is difficult to see how any such puzzlement as that which besets Quine in the environment of (3) can arise.

There is, however, a serious downside to the current way of fixing Quine's theory of predication. Anyone thinking of predication as a kind of logical form (or "construction", if one adheres to Quine's lingo) is bound to be upset by the realization that in many cases what qualifies as a predication will depend on the empirical facts. That the logical form of (3) should depend on the recently hypothesized but unverified Vulcan living up to its name no self-respecting Fregeian or Russellian would tolerate. Were 'Vulcan' never to name an existent planet, this would not in the slightest affect the status of 'Vulcan rotates' as a predication. Matters of logical form have no more to do with the empirical facts than has the status of 'Vulcan' as a singular term.

One can see the attraction of the Fregeian and Russellian ways of looking at predication in predictable responses to the current method of resolving the problem in Quine's theory of predication. Russell held virtually all atomic sentences (containing what Quine, but not Russell, would call singular terms) to be non-predications, except perhaps those with the demonstratives 'this' and 'that' in subject position.

---

[18] Ibid., p. 199.

And Frege, at least in his scientific mood, regarded them all as predications by arbitrarily stipulating a referent for singular terms such as 'Vulcan'. Thus the status of a sentence as a predication would not depend on the empirical facts of reference in either case.

There is another way of visualizing predication that salvages the extensional components of Quine's theory of predication, a way not subject to the complaint that it makes being a predication sometimes dependent on a factual outcome.[19] It too stems from the intuition that predications are singular sentences such that *were* their constituent singular terms to refer, their truth (or falsity) *would* depend on whether the constituent general term is true (or false) of the referents n-wise of those singular terms. Notice that were (3) counted false or truth-valueless, it could still qualify as a predication according to the basic intuition

Quine's view of the logical form (or "construction") called predication, like Kaplan's general view of logical form, is a semantical view; the logical form of a sentence depends on the way in which it is evaluated for truth-value. So, adopting the Quinian style of talk, another way of realizing the intuition expressed in the preceding subjunctive is to say that: *predication* joins a general term to singular terms to form a sentence that is true according as there exist objects n-wise to which the singular terms refer and of which the general term is true a, and otherwise is false.[20] 'Vulcan is an object **x** such that **x** rotates' is now trivially false but nevertheless qualifies as a predication. (Of course, the evaluation rule now defining predication also makes

(5)   The Sun is an object **x** such that **x** rotates

false because 'is an object **x** such that **x** rotates' is false of the referent of 'the Sun'.) Most importantly, substitution of the predicate

(is an) object **x** such that **x** exists $\supset$ **x** is identical with **x**

in the false (3) yields

---

[19] The formal counterpart of the doctrine to follow is present already in Ronald Scales' Ph.D. Dissertation, *Attribution and Existence* (University of California, Irvine), University of Michigan Microfilms (1969). Scales develops a version of negative free logic including complex predicates in which a sentence of the form **t** is **P** is not always logically equivalent to **t** (is such that it) is **P**. See also the last section of Chapter 7.

[20] This view of predication is certainly harmonious with the developments in the chapter on regimentation in Quine's *Word and Object*.

(6)   Vulcan is an object **x** such that **x** exists $\supset$ **x** is identical with **x**

But (6) no longer implies the true

(3*)   Vulcan exists $\supset$ Vulcan is identical with Vulcan

because the classical rule of *concretion* holds in Scales' treatment only in the form

   (Con) If **t** (is such that **it**) is **P**, then **t** exists and **t** is **P**.

(Con) makes it easy to see why the current revision of Quine's theory of preserves *salva veritate* substitution of co-extensive predicates. Replacing 'is an object **x** such that **x** rotates' in (3) by the co-extensive predicate 'is an object **x** such that **x** rotates if it exists' yields

(6)   Vulcan is such that it rotates if it exists.

But (6) no longer is true; given **(Con)**, it is equivalent to the false

(7)   Vulcan exists and (if Vulcan exists, then Vulcan rotates).

Thus does the revised theory of predication avoid the "stain" of non-extensionality *vis a vis* predicates (general terms). And since substitutivity of co-referential singular terms still holds, the extensionalistically inclined philosopher who does not think it wise, even if possible, to dispense with singular terms is no longer left gasping for air. In the current revision of Quine's theory, the way the world turns out determines the truth-value of a statement but not its logical form.

It is useful to compare this account of predication with Russell's. The very same sentences the present revision counts as predications, Russell would deny that status. Nevertheless in both theories these sentences get the same truth-value. Moreover, his theory of scope also survives albeit as a function of the connectives and other logical operators involved rather than of the singular term, (or, in his case, definite descriptions). Consider, for example, the sentence

(8)   Vulcan is a nonexistent.

This is ambiguous. It could mean the false "negative" predication

(9)   Vulcan is an object **x** such that **x** does not exist

or it could mean the true negation of the "positive" predication

(10)   It is not the case that Vulcan is an object **x** such that **x** exists.

The ability to accommodate scope considerations a la Russell is, of course, dependent on the 'such that' idiom.

The current repair of Quine's theory of predication preserves the association in unregimented discourse between referential transparency and predication, a hallmark of Quine's theory of referential opacity. The problems associated with predications involving singular terms without existential import are resolved with the exception perhaps of the alleged anomaly ensuing from admitting such singular terms to purely referential positions in predications. But the anomaly evaporates with the insertion of the words 'if any' in the appropriate place in the statement of motivation underlying the concept of purely referential position. Of course, the problems with predication involving singular terms without existential import also evaporate in regimented discourse given the controversial elimination of all singular terms except variables. What the above repair suggests, however, is that the transition from unregimented discourse to regimented discourse *vis-à-vis* singular terms need not be so radical as Quine seems to have thought at least convenient and at most essential.

7

Nonextensionality

1. INTRODUCTION

Consider the statement

(1) Necessarily $9 = 9$.

(1) is believed by most philosophical logicians to be non-extensional according to each of two prominent conceptions of extensionality. First, it fails the *salva veritate* substitution test when '9' is replace by the co-referential singular term 'the number of planets'. Second, in possible world semantics, its truth is regarded as dependent on the reference-in-all-possible-worlds of '9', or, alterntively, on the intension of '9' conceived as a function from possible worlds to individuals.

More generally, a statement is **SV-Extensional** according to **the** *salva veritate* **substitution conception** just in case singular terms co-referential with a statement's constituent singular terms, predicates co-extensive with the statement's constituent predicates, and statements co-valent with a statement's constituent statement(s), substitute in that statement *salva veritate*. A statement is **not SV-Extensional** if there is at least one failure of *salva veritate* substitution. (This is close to the characterization in Quine's *Word and Object*.[1])

---

[1] Willard Van Orman Quine, *Word and Object*, MIT Press, Cambridge and Wiley, New York (1960), p. 151. Quine uses the expression 'general term' rather than 'predicate', a difference which is important later in this essay.

A statement is **TD-Extensional** according to **the truth-value dependence conception** just in case its truth value depends only on the extensions *simpliciter*, if any at all, of its constituent singular terms, predicates and statements. A statement is **not TD-Extensional** if entities other than extensional entities are involved in the computation of its truth value.

That a statement may be nonextensional according to the truth value dependence conception but not according to the *salva veritate* conception is evident in Frege's treatment of indirect *(ungerade)* discourse. For instance, Frege, who disallowed failures of substitutivity of identity, nevertheless held that the truth value of statements such as

(2)   The man who looked like Beethoven but acted like Wagner was believed by Peter to be a philosopher,

depended in part on the *sense* of 'the man who looked like Beethoven but acted like Wagner', presumably an intensional entity. So (2) is nonextensional according to the truth value dependence standard, but may not be accordng to the *salva veritate* substitution conception; the referent of 'the man who looked like Beethoven but acted like Wagner' is a certain *sense* in (2) but not (for Frege) in the statement

(3)   The man who looked like Beethoven but acted like Wagner is the man who wrote *Fuzzy Thoughts, Clear Beliefs,*

which is true – take my word for it. So, even though (3) guarantees the co-referentiality of its constituent singular terms, there can be no violation of *salva veritate* substitution *vis-à-vis* the singular terms in (2) because in (3) they pick out the same individual rather than the same sense. What is not so often recognized, however, is that the relationship – or rather the lack thereof – also holds the other way round; a statement can be nonextensonal in the *salva veritate* substitution sense, but be extensional in the truth value dependence sense.

It may seem remarkable, and perhaps even repugnant to many, that a statement can fail the *salva veritate* substitution test, yet its truth value be dependent solely on the extensions *simpliciter*, if any at all, of its constituent singular terms, predicates, and statements. But with the advent of free logics, such a result is easily demonstrated. The general argument for this claim has been well established for almost three decades.[2] It will suffice here to present merely a sketch.

A free logic is a predicate logic allowing singular terms having no existential import but whose quantifiers nevertheless retain the interpretation in classical predicate logic. Assume such a logic containing *only* identities as its atomic statements. In such logics,

(4)   **t** exists, where **t** is a singular term constant,

is shorthand for

(5) $\exists x(x = t)$.

So it is convenient but not necessary to use the locution (4) in what follows.[3]

Suppose also that the universe of discourse contains (intuitively) only existent objects, if any objects at all, and that the denotation function defined on the singular terms is partial. In such model theoretically based free logics, the predicates 'is self-identical', 'is a self-identical existent' and 'is self-identical if existent' are coextensive.

Now consider the statement

(6) $1/0 = 1/0$.

(For convenience, read (6) as '1/0 is self-identical'). No matter what truth-value this statement is assumed to have, including none at all, *salva veritate* substitution of coextensive predicates will fail. For example, if (6) is true (as in most *positive* free logics), substitution of 'is a self-identical existent' for 'is self-identical' in the host statement yields the statement

(7) $1/0 = 1/0$ **&** $1/0$ exists.

---

[2] See Chapter 6.

[3] In the late 1950s, Hintikka proved that in a certain free logic singular existence was justfiably definable in tems of existential quantification and identity. In the 1960s, I showed that Hintikka's result carried over to any free logic. In 1982, Meyer, Bencivenga and I proved that in a free logic consisting of the usual truth functional connectives, quantifiers, singular terms and general terms, but not identity, singular existence was not definable. (See Chapter 8.) In 1986, Bencivenga and I showed that if a free logic sans identity is augmented with a complex predicate forming operator, singular existence could be defined in a manner reminiscent of an informal suggestion by Arthur Prior. (See Karel Lambert and Ermanno Bencivenga, 'A free logic with simple and complex predicates', *Notre Dame Journal of Formal Logic*, 27 (1986), pp. 379–392.)

(Read: '1/0 is a self-identical existent'). But (7) is false. Similarly, if the host statement is taken to be false (as in *negative* free logics), and 'is self-identical' is replaced by 'is self-identical if existent', statement

(8) $1/0$ exists $\supset 1/0 = 1/0$

(read: '1/0 is self-identical if existent') results. But (8) is true. And so on. (See Chapter 6.) So even in as sparse a language as imagined above, a language containing no modalities, psychological verbs, or the like, statements such as (6) are nonextensional in the *salva vritate* sense.

Consider now that species of free logic called *negative free logic*. In these logics *all* atomic statements containing at least one singular term having no existential import are false. For example, (6) would be false because '1/0' refers to no existent, and hence has no extension, given the supposition above about the make-up of the universe of discourse and the nature of the denotation function. So the computation of its truth-value depends only on the extensions, if any, of the singular term '1/0', and the predicate 'is self-identical' (assuming identity is not regarded as a logical constant). Since this is the semantical policy in most developments of negative free logic, (6) is extensional in such logics, according to the truth value dependence conception.

Since the early 1980s, mechanistic mathematicians have developed theories of partial functions founded in fact on free logics for use in the development of computer programing languages. An example is Farmer's IMPS[4], which is based on negative free logic; another is Dana Scott's COLDS[5] which is based on a positive free logic, a perhaps more familiar species that counts *some* statements containing only singular terms not having existential import true – (6), for example. As emphasized by van Fraassen and me in 1972[6], free logics recommend themselves as a natural foundation for the theory of partial functions. These logics regard names such as '1/0' as genuine singular terms (contra classical predicate logic) even though the function specified – division,

---

[4] William Farmer, 'A simple theory of types with partial functions and subtypes', The MITRE Corporation, Bedford, MA 137.

[5] L. M. G. Feijs and H. B. M. Jonkers, *Formal Specification and Design* (Cambridge Tracts in Theoretical Computer Science 35), Cambridge University Press, Cambridge (1992).

[6] Karel Lambert and Bas van Fraassen, *Derivation and Counterexample*, Dickenson, Encino, CA (1972).

for example – may fail to have values for some pairs of arguments; the pair <1,0>, for instance.

Because the underlying logics in recent treatments of partial functions essential to programming theories like IMPS and COLD are free, the quesion arises whether the failure of the *salva veritate* substitution of co-extensive predicates in such logics cannot be undone or at least explained away. Such an argument or explanation would amount to an attempt to subvert, in one way or another the demonstration above against the *salva veritate* substitution of co-extensive predicates in free logic, and, thus to an attempt to subvert the argument that extensionality in the sense of truth value dependence is independen of extensionality in the sense of *salva veritate* substitution.

In the remainder of this essay, two possible attempts at subversion will be critically examined. One attempt concerns the notion of 'predicate', and depends on Henry Leonard's treatment of predicates in his essay 'Essences, attributes and predicates'.[7] The other attempt concerns the notion of co-extensive, and is latent in the inner domain-outer domain semantics for free logic (especially in Scott)[8].

2. CO-EXTENSIVE PREDICATES REVISITED

A critic of the master argument is sure to point out that it depends entirely on the kind of model structure presumed in a free logic; only when the models are of the sort described above, in which the single universe of discourse is imagined to consist solely of existent objects, does the master argument succeed. For only then will the predicates 'self-identical', 'is a self-identical existent' and 'is self-identical if existent' be co-extensive. If, however, the universe of discourse is of the inner domain-outer domain variety where only the inner domain is imagined to comprise the existents, the master argument fails. The predicates in question will no longer be co-extensive; in particular, the set of things that are self-identical will be distinct from the set of things that exist and are self-identical.

[7] Henry S. Leonard, 'Essences, attributes and predicates', *Proceedings and Addresses of the American Philosophical Association*, 37 (1964), pp. 25–51.

[8] Dana Scott, 'Existence and description in formal logic' in *Bertrand Russell: Philosopher of the Century* (ed. R. Schoenman), London, Allan & Unwin) (1967), pp. 181–200.

Thus, imagine only the inner domain to consist of existent objects and the outer domain to consist of non-existents, for example as in Scott's semantical theory of virtual classes.[9] Imagine the interpretation function defined on the singular terms and general terms to be total. Then every singular term gets assigned a member in the union of the inner and outer domains, and every n-adic predicate gets assigned a set of n-tuples of members from the union of the inner and outer domain – the union of the existents and non-existents. Then the predicate 'is self-identical' will have in its extension virtual entities like the Russell set (or Mill's round square), but the extension of the predicate 'is a self-identical existent' will contain no such "entities." So the two predicates cannot be co-extensive, and *salva veritate* substitution of co-extensive predicates is peserved. So, it may be urged, if *salva veritate* substitution of co-extensive predicates is adopted as a requirement for free logics, then model structures in which the universe of discourse is of the inner domain-outer domain kind ought to be preferred.

This way out of the master argument, appealing though it may be at first glance, is not cost free. The costs to be detailed here are different frm the usual complaints about the mystery and/or incomprehensibility of the notion of non-existent (or virtual) objects. Indeed, much of the sting of this complaint has been removed by the provably sound accounts of Parsons [1980], Pasniczek [1988] and others.[10]

For starters, consider the notion of a partial function, and, in particular, the two place partial function of division. Given the sort of model structure outlined in the previous paragraph, this function either no longer is partial or is partial only with respect to a subset of its range. For '1/0' now turns out to have a value, though to be sure not a value among the existents. This is a bizarre solution. It is like adopting an explanation of the behavior of the man who looks like Beethoven but behaves like Wagner that requires denying that there is such a person or holding that the explanation instead applies to his cousin.

In the second place, the current objection succeeds at the expense of obliterating the traditional notion of extension. The traditional

---

[9] Dana Scott, 'Existence and description in formal logic', in *Bertrand Russell: Philosopher of the Century*, (ed. R. Schoenman), Allan & Unwin, London, pp. 181–200.

[10] Terence Parsons, *Nonexistent Objects*, Yale University Press, New Haven (1980); Jacek Pasniczek, *The Meinongian Version of Classical Logic*, Lublin, Wydawnictwo Uniwersytetu Marii Curie-Sklodowskiej (1988).

noion is that the extension of a predicate is the set of *existent* entities of which that predicate is true (or false). What passes for the extension of a predicate nowadays more closely resembles what C. I. Lewis called its *comprehension*. That is, it is the set of entities existent or nonexistent of which the predicate is true (or false). The notion of comprehension so characterized is an important one. For example, it is important in the distinction between natural and strict modality (and hence to the explanation of the sense in which the statement 'Nothing is both simultaneously red and green all over' is true "by virtue of the meanings of words alone").[11] But if the traditional notion of extension is re-introduced, the master argument against the principle that co-extensive predicates substitute *salva veritate* again prevails because the predicates used in that argument *are* co-extensive in the traditional sense of the word 'co-extensive'.

In the third place, even if one cleaves to the expanded sense of 'extension', unless the outer domain consists of a limited, finite set of objects, there can be no object language counterpart of co-extensive predicates without doing something seemingly quite alien to the very enterprise of free logic. To fix the ideas, imagine the simplest case in which the outer domain consists of a single member. Then the object language measure of co-extensive predicates would be the truth of

(9) $\forall x(Px \equiv Qx) \ \& \ (P_* \equiv Q_*),$

where P and Q are predicates, and '$*$' is the name of the single entity in the outer domain of the universe of discourse. This method of "expressing" the notion of co-extensive predicates in the object language can be extended, if the number of entities in the outer domain is finite just by adding the appropriate finite number of conjuncts to the conjunctions of biconditionals in (9). But what happens if the outer domain is infinite as is the case in Scott's theory of virtual classes? The number of conjuncts that have to be added must be infinite, thus precluding an object language counterpart of what it means for predicates to be co-extensive on the model of (9). The obvious resolution would be the addition of another quantifier intended to range

---

[11] Karel Lambert and Bas van Fraassen, 'Meaning relations, possible objects and possible worlds', in *Philosophical Problems in Logic*, (ed., Karel Lambert) Reidel, Dordrecht (1970), pp. 1–20.

over the members in the union of the inner and outer domains. For
instance, on might now introduce the "wider" universal quantifier

[∀x]

(to be read: 'Everything x, existing or not') and "express" co-extensive
predicates in the object language by

(10)   [∀x](P(x) ≡ Q(x)).

The suggestion is quite alien to free logic which recognizes only ac-
tualist quantifiers as basic. Indeed, with the wider quantifier [∀x], the
quantifiers of free logic would now be *definable*, given an existence
predicate; for instance,

∀x(A)

would simply be an abbreviation for

[∀x](E!(x) ⊃ A).

Why not recognize an outer quantifier and regard free logics, formally
speaking, just as a subspecies of the wider predicate logic? Scott himself
has provided an answer. Conceiving the outer domain to consist of
virtual objects, he considers the question whether the virtual objects
should be quantified over and replies:

I think the answer should be a firm no. If we have come to value the virtual entities so highly that we want to quantify over them, then we have
passed to a *new* theory with a *new* ontology (and with new virtuals also!).
The role of virtual entities is to make clear the structure of the basic
domain D [the inner domain], not to introduce a whole new collection of
problems. That is why we are happier when references to virtuals can be
eliminated.[12]

It might be replied that this sort of objection to a quantifier over everything does not apply to other semantical developments of free logic
in which the outer domain consists of expressions as in the nominal
domain-real domain approach by Robert Meyer and me.[13] There an

---

[12] Dana Scott, 'Advice on modal logic' in *Philosophical Problems in Logic* (ed. Karel
Lambert), Reidel, Dordrecht (1969), p. 146–147.

[13] Robert Meyer and Karel Lambert, 'Universally free logic and standard quantification
theory,' *Journal of Symbolic Logic*, 33 (1968), pp. 8–26.

outer quantifier would range over existent things, namely, concrete expressions. Unfortunately, this rebuttal is unsatisfying in two respects. First, the extension of a predicate now consists of things and words; for example, the extension of '__ is a horse' consists not only of actual horses but also of the expression 'Pegasus' (rather than the nonexistent but possible horse captured by Bellerophon). Despite certain possible technical advantages of this second intention-like approach, it must be admitted that it is not philosophically very satisfying. Second, even if natural repugnance to this way of viewing the contents of the extension of any predicate could be overcome, there is the difficulty pointed out by Bencivenga[14] that this approach to the semantics of free logic requires a prior *primitive* classification of singular terms into those which refer and those which don't, a classification which cannot be represented in the formal language itself.

A final compelling objection is this. It will always be possible to introduce a predicate Φ whose extension can be taken to be the widest possible domain, and a singular term, say,

ix(Px & ~Px)

having no referent even in the widest domain. Intuitively,

Φ ix(Px & ~Px)

will be false. But then failure of the *salva veritate* substitution of co-extensive predicates can be resurrected merely with the predicate 'is self-identical', no matter how widely the quantifiers are construed.

3. PREDICATES AND GENERAL TERMS

Another attempt to undermine the argument that *salva veritate* substitution of co-extensive predicates fails in free logic turns on narrowing the answer to the question, "What is a predicate?". Following a tradition beginning at least with Peirce and reaffirmed by Leonard,[15] a sharp distinction is made between expressions like '__ is a planet'

---

[14] Ermanno Bencivenga, 'Free logics', in *Handbook of Philosophical Logic, Vol. III* (eds. Dov Gabbay, et al.), Reidel, Dordrecht, 1986, pp. 373–426.

[15] See footnote 7.

and general terms like 'object such that it is a planet'. Predicates are exclusively *identified* with general terms, and hence as expressions that are true (or false) of each of possibly many things in a given class. (Peirce called statement fragments such as '__ is a planet' *rhemes*.)

The view just described is a departure from the conception of predication in conventional applications of modern predicate logic to colloquial discourse, a conception that views a predicate as what is obtained by deleting at least one singular term from a statement. In more traditional language, general terms are *categorematic*, but rhemes are only *quasi-categorematic* because they are not the sort of expression that even purports to be true or false of anything.

General terms, however, can be generated out of rhemes in the following way. From '__ is a planet', for example, open sentences can be obtained by replacing the gap with a free variable, and general terms – hence predicates – can be generated from them by prefixing to them a variable binding general term forming operator, here designated by

Δ.

For instance, prefixing

x is a planet

with

Δ **x**

yields the general term

Δ**x**(**x** is a planet)[16]

and may be read as

object **x such that x** is a planet.

And the same applies to rhemes with more than one gap – relational rhemes. It now becomes possible to discriminate between statements like

---

[16] I prefer this sign over λ to avoid confusion between singular and general terms that Church's famous sign tends to promote.

(11)   Vulcan is a nonexistent,

that is,

(12)   Vulcan is an object **x such that** it is not the case that **x exists**,

a predication involving a "negative" general term, and statements like

(13)   It is not the case that Vulcan exists,

that is,

(14)   It is not the case that Vulcan is an object **x such that x** exists,

the negation of a predication involving a "positive" general term. Significantly, for the negative free logician, (12) is false, but (13) is true following the spirit if not the letter of Russell. This necessitates the following restricted form of general term abstraction:

(14) $\Delta \mathbf{x} \mathbf{A}, \mathbf{t} \equiv (\mathbf{t}$ exists $\&\ \mathbf{A}(\mathbf{t}/\mathbf{x}))$,

that is,

(15)   **t** is an object **x such that A(x)** if and only if both **t** exists **and A(t/x)**, where **t** is an arbitrary singular term constant, and **A(x)** is an arbitrary rheme containing the free variable **x**.

Now compare

(16)   Vulcan exists $\supset$ Vulcan is identical with Vulcan,

and

(17)   Vulcan is an object **x such that** $(\mathbf{x}$ exists $\supset \mathbf{x} = \mathbf{x})$.

(16) is true in negative free logics, but (17) is false (in virtue of (14)). It would appear then that if 'predicate' is narrowly construed in the sense of Leonard (and perhaps Peirce), the failure of *salva veritate* substitution of co-extensive predicates can be avoided after all. For in negative free logics, the predicates (general terms)

    object **x such that x** is identical with **x**

and

    object **x such that** if **x** exists, then **x** = **x**

are co-extensive because

(18) $\forall y((y$ is an object **x such that x** $= x) \equiv (y$ is an object **x such that** **(x** exists $\supset$ **x** $= x)))$

is true. But both

(19) Vulcan is an object **x such that x** $= x$

and (17) are false given (14). Indeed, this test of extensionality conforms exactly to Quine's own wording on page 151 of his *Word and
Object*, where *salva veritate* substitution concerns statements and *terms*,
singular and general. Moreover, the importance of the emphasis on
general terms looms even larger in view of the fact that in negative
free logic the inference from

(20) $\forall x(x$ is identical with **x** $\equiv$ **(x** exist $\supset$ **x** $= x))$

and (17) to

(21) Vulcan is identical with Vulcan

is *invalid'*.

The philosophical idea behind the technical distinction between
rhemes and general terms is easily expressed. It is that only statements
like (12), (14), (17) and (19) concern predication; indeed, this again
conforms closely with Quine's conception of predication in the unregimented language of *Word and Object* in which predication is defined
not by means of the expression 'predicate' but rather by means of the
expression 'general term'. Nevertheless, there are serious questions
with this way of undermining the argument that in free logic *salva
veritate* substitution of co-extensive predicates fails, questions having
to do both with the significance of the distinction between rhemes
and predicates, and with the motivation underlying that distinction.

In the first place, unless rhemes are treated as a kind of quasi-
predicate at least in the sense of having extensions it is difficult to
see how a semantical account of the difference in truth value between
(19) and (21), let alone of the difference in truth-value between (16)
and (17), would proceed. In fact, rhemes are treated in just this way
in most semantical developments of free logic having a term forming
operator like Δ. But then to say that negative free logics do not violate

the *salva veritate* substitution of co-extensive predicates seems arbitrary turning on a semantically inessential feature of expressions of the form '__ is a planet'. For, as the argument from (20) and (17) to (21) shows, in negative free logic, co-extensive rhemes do not substitute *salva veritate*.[17]

In the second place, the belief that only predications imply the existence of the purported references of their constituent singular terms in contrast to statements composed of rhemes and singular terms is not even shared by all proponents of the view that only statements of the form in (22) express predications. Leonard, for example, holds that whether a predication implies the existence of its purported references depends entirely on the character of the constituent predicate, whether, in his words, it is "existence entailing"; for Leonard the predicate 'object **x such that x** is fictional' is not an existence entailing predicate and the predication

(22) Pegasus is an object **x such that x** is fictional

does not imply the existence of Pegasus.[18] This sort of consideration again tends to weaken the motivation underlying the predicate-rheme distinction, and thus the credibility of the current attempt to undermine the master argument that *salva veritate* substitution of co-extensive predicates fails in many free logics.

Still, I see hope in this way out of the problem. For what it suggests is that there is fundamental difference between what logicians call open sentences and predicates. SV-Extensionality, after all, has to do with co-extensive predicates (or general terms), not with open sentences. It would be correct to say that both the sentences 'It is delightful' and 'That piece of music is delightful' contained the predicate 'is delightful'. It would be incorrect to say that the sentence 'It is delightful' is merely a stylistic variant of the predicate 'is delightful', and hence has the same extension. The predicate 'is delightful' has an extension all right, namely, the set of the things that are delightful. But the sentence 'It is delightful' has no extension at all until the extensions

---

[17] It is important to notice that the way of restoring extensionality *à la* Scales is quite different from the proposal here. In Scales case, no distinction is made between 'is self-identical' and 'object **x such that x** is self-identical'. Both qualify as predicates in Scales, though the latter is a complex predicate. See the last section of Chapter 6.

[18] Op. cit., Henry Leonard, 'Essences, attributes and predicates' (1964).

of its predicate and its pronomial subject 'It' – the counterpart of a variable – are determined. Then its extension would be a truth-value, if any at all, not a set.

Consider again statement (17), that is,
Vulcan = Vulcan.

Given the truth of (20), the statement in question *apparently* yields, by substitution of coextensive predicates, (21), that is,

Vulcan exists ⊃ Vulcan = Vulcan.

If (17) is false (as in negative free logics), then (21) is true, and the principle of the *salva veritate* substitution of coextensive predicates seemingly fails. But this is not quite correct, though the replacement in (17) of '__ is self identical' by '__ is self identical if existent' certainly suggests otherwise. Actually, what is shown by the example is a *consequence* of the substitution of coextensive predicates in which (17) is presumed to be false and (21) is true. Hence, it is concluded erroneously that the substitution principle in question fails. To get from (17) to (21) one must first infer, by the logical rule of *predicate abstraction*,[19]

(23)   Δx(x is identical with x), Vulcan

from (17). By *salva veritate* substitution of coextensive predicates, (23) yields

(24)   Δx(x exists ⊃ x is identical with x), Vulcan.

This, in turn, yields by the rule of *predicate concretion*,

(25)   Vulcan exists ⊃ Vulcan is identical with Vulcan.

But (24) and (25) differ in truth value only if one has available the *abstraction prinicple*

(25) $\Delta x(A), t \equiv A(t/x)$.

However, (25) doesn't hold in Scales' treatment of negative free logic (nor in the treatment of positive free logic with simple and

---

[19] See W. V. Quine, *Methods of Logic* (4th ed.), Harvard University Press: Cambridge, MA (1982), p. 134 ff.

complex predicates by Bencivenga and me)[20]. The upshot is that SV-extensionality need not fail in free logic. Evidently, then, the connection expressed in the conditional that if a statement if TD-Extensional it is also SV-Extensional can be reestablished even in languages containing singular terms without existential import.

---

[20] Karel Lambert and Ermanno Bencivenga, 'A free logic with simple and complex predicates', *Notre Dame Journal of Formal Logic*, 27 (1986), pp. 247–256.

8

The Philosophical Foundations of Free Logic

1. WHAT FREE LOGIC IS AND WHAT IT ISN'T

On page 149 of the second edition of their book, *Deductive Logic*, Hugues Leblanc and William Wisdom[1] say the following about the origins of free logic.

> Presupposition-free logic (known for short *as free logic*) grew out of two papers published simultaneously: Hintikka's 'Existential Presuppositions and Existential Commitments', *The Journal of Philosophy*. Vol. 56. 1959. pp. 125–137, and Leblanc and (Theodore) Hailperin's 'Nondesignating Singular Terms'. *The Philosophical Review*. Vol. 68, 1959. pp. 129–136. Both made use of the identity sign '='. In a later paper Karel Lambert devised a free logic without '=' ... (See Lambert's 'Existential Import Revisited', *Notre Dame Journal of Formal Logic*, vol. 4, (1963), pp. 288–292).

This account is in one respect misleading, and in another inaccurate. It is misleading because Rolf Schock, independently either of Hintikka or of Leblanc and Hailperin, was developing a version of free logic in the early 1960s quite different in character from those mentioned by Leblanc and Wisdom. Moreover Schock's ideas were in certain respects more fully developed because he also supplied models for his own systematization. The writers mentioned by Leblanc and Wisdom had suggested – in print at any rate – only informally the semantical bases for their formulations. (Apparently Schock's ideas were known to some

---

[1] Hugues Leblanc and William Wisdom, *Deductive Logic*, Allyn & Bacon, Boston (1976).

European scholars in the early 1960s but oddly were not disseminated. One can find an account of Schock's pioneering efforts in his 1968 book, *Logics Without Existence Assumptions.*[2])

The inaccuracy in the Leblanc-Wisdom remark concerns the failure to mention the earlier essay published in 1956 by Henry Leonard entitled 'The Logic of Existence'[3]; Leonard's essay, as a matter of fact, is cited in the 1959 paper by Leblanc and Hailperin. To be sure, Leonard's formulation was not as thoroughly developed as those mentioned by Leblanc and Wisdom, but then neither were the latter in comparison with Schock's later and independent work. So the key ideas were *first* enunciated in Leonard's paper, and it is his essay that is the direct and indirect source of influence for many if not most workers in free logic. It is appropriate, therefore, to step back now and see what he precipitated.

What is a free logic? The expression 'free logic' is an abbreviation for the phrase 'free of existence assumptions with respect to its terms, general and *singular*'. When the expression 'free logic' was coined in 1960, the motive was simply convenience and nothing more. It has become a widely adopted convention. It has also become the subject of lament even from pioneers in the subject. Hintikka, for example, objected to the choice of words on the ground that the use of the word 'free' in the expression 'free logic' was apt to be confused with the use of the word 'free' in the expression 'free algebra'.[4] Those of you have wondered why Petroleum V. Nasby's work, *The Free Love Society*, never received the audience it was due now have the explanation! Let me here acknowledge full responsibility for the expression 'free logic', and apologize for whatever unfortunate associations it may have encouraged.

What the expression 'free logic' abbreviates needs fuller explanation. Consider, first, the expressions 'singular term' and 'general term'. They are used here in the contemporary way. Quine has put the distinction this way. A *general term* is one true of each object or pair of objects or triple of objects, and so on, in a given class if of any thing

---

[2] Rolf Schock, *Logics Without Existence Assumptions*, Almqvist & Wiksells, Stockholm (1968).

[3] Henry Leonard, 'The logic of existence', *Philosophical Studies*, 7 (1956), pp. 49–64.

[4] Jaakko Hintikka 'Existential presuppositions and uniqueness presuppositions', in *Philosophical Problems in Logic* (ed. Karel Lambert) Reidel, Dordrecht (1970), p. 20.

at all; a *singular term* is one that purports to refer to just one object.[5] Examples of general terms are expressions such as 'man', 'talarias', 'greater than', 'brakeless trains', 'unicorn', 'satellite of the Earth' and 'between California and the deep blue sea'. Some of these expressions, for example, 'unicorn' and 'talarias' are true of no existing thing, and, thus, are said to have *no existential import*. Examples of singular terms are expressions such as 'the premier of Austria in 1980', 'suavity', 'being notable', 'Vulcan', 'Heimdal', '1/0', 'the man born simultaneously of nine jotun maidens',[6] and 'Potsdorf'. Some of these singular terms, for example, 'Vulcan', 'Heimdal', '1/0', 'the man born simultaneously of nine Jotun maidens', and 'Potsdorf' refer to no existing thing, and thus have no existential import.

The explanation of the phrase logic free of existence assumptions with respect to its terms, general and singular', then, is this. It is a logic in which the quantificational phrases 'every' and 'some', and their stylistic variants, have their classical interpretation and there are no statements that are logically true only if it is true that **G** exists for all general terms **G**, or it is true that **s** exists for all singular terms **s**. It is this explanation that really *defines* the expression 'free logic'.

To get some perspective on the explanation, consider, first, a logic *not* free of existence assumptions with respect to its general terms. For instance, consider a logic including the statement

(1)   If all men are mortal, then there exist men that are mortal

as logically true. Notice that the statement 'Round squares exist' – that is, 'There exists round squares' is false. If, however, the general terms 'men' and 'mortal' are replaced by the general term 'round squares' the resulting statement

(2)   If all round squares are round squares, then there exist round squares that are round squares

---

[5] W. V. Quine, *Word and Object.* Wiley, New York (1960), p. 96.
[6] Heimdal, indeed, is the man born simultaneously of nine sibling Jotun maidens. The above examples are all *constant singular terms.* In most formulations of elementary logic there occur individual variables, such as the letters **x**, **y**, and **z**, and sometimes also property variables such as the letters **P**, **Q**, and **R**. Both kinds are usually treated as a species of singular terms, that is, as *variable singular terms.*

is not true. So, to preserve the logical truth of the original statement (1), the general term 'round squares' must not be allowed to replace the general terms 'men' and 'mortal' in the specimen statement. In general, no general term falsifying the condition

G exists

can be allowed to replace the general terms 'men' and 'mortal' in (1). So the logic containing that statement among its logical truths cannot be free of existence assumptions with respect to its general terms. Why? Because there are statements in that logic that will not be logically true *unless* it is true that **G** exists for any general term **G**.

Consider next a logic that is not free of existence assumptions with respect to its singular terms. In particular, consider one in which the statement

(3)   There exists something that is identical with Bush

is logically true. Notice that the statement '1/0 exists' is false, and that if the singular term '1/0' is put in the place of the singular term 'Bush' in (3), the result is the untrue statement

(4)   There exists something that is identical with 1/0.

So to preserve the logical truth of (4) the singular term '1/0' must not be allowed to replace the word 'Bush' in (3). In general, no singular term that falsifies the condition

s exists,

can be allowed to replace the singular term 'Bush' in (3). So a logic containing that statement among its logical truths is not free of existence assumptions with respect to its singular terms because, again, there are statements in it that will not be logically true unless it is true that **s** exists for any singular term **s**.

Logics that are free from existence assumptions with respect to their general and singular terms are now in plentiful supply. There are many others in addition to the logics mentioned by Leblanc and Wisdom including apparently some versions of Lesniewski's "Ontology" developed during the 1930s.[7]

---

[7] See Karel Lambert and Thomas Scharle, 'A translation theorem for two systems of free logic', *Logique et Analyse*, 40 (1967), pp. 328–341.

There is an important implication of the preceding remarks to which attention should be drawn. As suggested a moment ago, in a free logic with identity, the statement

(5)   There exists something identical with Heimdal

is not true. On the other hand, the statement

(6)   Every existing object is such that there exists something identical with it

is logically true in all free logics with identity. It follows that the logical principle known as *Universal Specification* fails, that is, the principle

(7)   If for every existing thing **x**, ...**x**... then ...**s**..., where **s** is a singular term, and if a variable, subject to the usual restrictions about the capturing of free variables,

Similarly, the corresponding inference rule known as *Universal Instantiation*, that is,

(8)   From every existing thing **x**, ...**x**... infer ...**s**..., where **s** is a singular term, and if a variable subject to the usual restrictions *vis-à-vis* the capturing of free variables,

is rejected in free logic.

Rejection of *Specification* (or *Universal Instantiation*) is usually what comes to mind when one sees the phrase 'free logic'. But caution is needed here. For though it is the case that rejection of the above principle (or rule of inference) is a necessary condition for a logic to be free, it is not a sufficient condition. The system of Soren Stenlund discussed in Chapter 4 is ample testimony to this fact.

Different kinds of free logic are possible according to the truth-values had by their simple statements. For example, suppose a language with the two-place predicate 'is identical with' as its only predicate, and all sorts of singular terms. Then the simple statements of that language might include statements such as '1/0 is identical with 1/0', 'Heimdal is identical with Heimdal', 'Heimdal is identical with Voltaire', 'Heimdal is identical with Pegasus', 'Potsdorf is identical with Salzburg', and so on. Now some free logicians, in the spirit of Russell, think all of the listed statements, indeed, *all* simple statements containing at least one singular term without existential import, are false. A free logic so motivated is a *negative free logic*.[8]

Other free logics count at least the statements '1/0 is identical with 1/0' and 'Heimdal is identical with Heimdal' true, and some, in the spirit of Frege's attitude toward scientific language, count *any* simple identity statement containing only singular terms that do not refer to existents true. Free logics of this sort are *positive free logics*.[9]

Finally, there are those who think all of the example simple identity statements are truth-valueless in the spirit of Frege's attitude toward colloquial discourse. Free logics counting all simple statements containing at least one singular term that does not refer to an existent truth-valueless (except, perhaps, simple statements of the form

s exists,

where **s** does not have existential import) are nonvalent *free logics*.[10]

Different opinions about the truth-values of simple identity statements containing singular terms that do not refer to existing objects help determine which statements are logically true. For example, only in positive free logics is the statement '1/0 is identical with 1/0' logically true. For that reason Descartes' *Cogito ergo sum*, construed as an argument, is invalid in positive free logics; 'If I think then I am' is false when 'think' is replaced with 'is identical with', and 'I' is replaced by a singular term without existential import. Again, only in negative free logics is the statement 'If I think, then I am' logically true (because 'I think' will always be false when 'I' is replaced by a singular term without existential import). Finally, in nonvalent free logics instances

---

[8] See, for example, Tyler Burge in 'Truth and singular terms', *Noûs*, 8 (1975), pp. 309–325.

[9] See, for example, Robert Meyer and Karel Lambert, 'Universally free logic and standard quantification theory', *Journal of Symbolic Logic*, 33 (1968), pp. 8–26. For an example of a positive free logic in the spirit of Frege, see Karel Lambert, 'A theory of definite descriptions' in *Philosophical Applications of Free Logic* (ed. Karel Lambert), Oxford, New York, 1991, pp. 17–28. In this development, the extensional principle

$(\sim E!s \ \& \ \sim E!t) \supset s = t$

is a theorem, and since the system is provably sound, also is logically true.

[10] See, for example, Scott Lehmann, 'Strict Fregean free logic', *Journal of Philosophical Logic*, 23 (1994), pp. 307–36. The well-known treatment of free logic by Bas van Fraassen based on supervaluations is not a nonvalent free logic despite the admission of truth-value gaps. In van Fraassen's treatment any statement of the form

s = **s**, where **s** is without existential import

is true. So it is a positive free logic.

of a version of the first order counterpart of the principle of the
nonidentity of discernibles might fail, that is, formally speaking,

ND   (G(s) & ~G(t)) ⊃ ~s = t.

For notice that intuitively both the statements 'Voltaire exists' and 'It
is not the case that Heimdal exists' are true. But the statement 'It
is not the case that Voltaire is identical with Heimdal' might be regarded as truth-valueless in virtue of the truth-valuelessness of the
simple statement 'Voltaire is identical with Heimdal'.[11] In view of
the above, it behooves those philosophers who lament *ad nauseum*
logic in general, and the "philosophically empty activity" of deciding on the truth values of simple sentences in particular, to be less
strident.[12]

   The preceding remarks sketch what a free logic is in the barest
of outlines. Since there are some misunderstandings and confusions
about it, some pretty well ingrained, it is appropriate to say what a free
logic is not. This discussion will help to make more vivid this particular
specimen in what Russell would have called the "logical zoo."

   Some have thought free logic to be presumptuous, that no logic is
free of all presuppositions. This complaint no doubt derives from the
unfortunate inclination of some free logicians, especially during the
formative years of the subject, to refer to free logic as 'presupposition-
free logic'. Still it is difficult to understand how this objection could
ever have had any credibility except to those whose inflexibility
matches the Gnat's in *Alice in* Wonderland.[13] For it is very clear from
even the most superficial inspection of presupposition-free logic that
the presuppositions of concern are *existence presuppositions*, and those
only.

---

[11] Considerations of this sort probably were among the reasons driving Skyrms to the
extremity of analyzing singular existence claims as metalinguistic assertions about
the referential status of their constituent singular terms. See Brian Skyrms, 'Supervaluations: identity, existence and individual concepts', *The Journal of Philosophy*, 64
(1968) pp. 477–483.

[12] As early as 1937, the logician C. H. Langford observed:
'Logic . . . is not in a particularly fortunate position. On the one hand, philosophers
prefer to speak of it without using it, while on the other hand mathematicians prefer
to use it without speaking of it and even without desiring to hear it spoken of'. I am
grateful to Aldo Antonelli for bringing this remark to my attention.

[13] It was the Gnat who exclaimed, 'What's the use of their having names if they won't
answer to them?'

An example of a logical presupposition not at issue in free logic, for instance, is the presupposition that the inference patterns both in standard and free logic require invariability in the senses of expressions occurring more than once in an argument. It is because of this precondition that the argument

> Everything is such that if it is a nut then it grows on trees.
> Something that is a nut has threads
> So, something that has threads grows on trees

is not accepted as evidence of the non-validating character of the inference pattern

> Everything is such that if it is (an) **F** then it is (a) **G**
> Something that is (an) **F** has **H**
> So, something that has **H** (is) **G**

despite the apparent truth of the premises and the falsity of the conclusion. The example argument is an instance of an equivocation, an argument that masquerades as a counterexample in virtue of a word or phrase – in this case the word 'nut' – having different senses at different occurrences in the argument. When the word is given the same sense throughout, the counterexample evaporates because one or the other of the premises will turn out false.

This is a good occasion to question the wisdom of using the word 'presupposition' at all in characterizing free logic. In recent philosophical logic, the word 'presupposition' has come to be associated with a conception of Frege's, later employed by Strawson in his critique of Russell's theory of definite descriptions and thoroughly studied semantically by van Fraassen, and many others.[14] The idea is that a statement **A** presupposes a statement **B** just in case whenever **A** is true, **B** is true and whenever it is not case that **A** is true, **B** is true. For example,

---

[14] Gottlob Frege, 'Über Sinn und Bedeutung', *Zeitschrift für Philosophie and philosophische Kritik,* Vol. 100 (1892), C (new ser., 1890), pp. 25–50; P. F. Strawson, 'On referring', *Mind,* LIX (1950), pp. 329–344; and Bas C. van Fraassen, 'Presupposition, implication and self-reference', *The Journal of Philosophy,* Vol. LXV (1968), pp. 136–152. Indeed, it was I who suggested to van Fraassen the possibility of using his notion of supervaluations to formalize the Frege-Strawson notion of presupposition, as he generously acknowledged in his essay.

some people believe the statement 'The present King of France is wise' presupposes the statement 'The present King of France exists' just because the former statement wouldn't be true or false unless the second statement were true.

In the first place, if one looks at the entire set of statements comprising a given logic, then some free logics are not free of existence presuppositions in the sense of the word 'presupposition' just explained. For example, in the formulation by van Fraassen, in 'Singular Terms, truth-value gaps and free logic',[15] when either a statement of the form '**s** is a planet' or 'it is not the case that **s** is a planet' is true, so is a statement of the form '**s** exists'. In the second place, if the focus is just on the set of logical truths, the relevant question now is not whether the *truth* of those principles depends on the truth of further statements of the form '**G** exists', for all general terms **G**, or of the form '**s** exists', for all singular terms **s**. Rather it is whether their *logical* truth so depends. In a free logic, as noted earlier, the answer is no, and hence free logics are free of existence presuppositions. Perhaps, to avoid 'confusion with the Frege-Strawson notion of 'presupposition', the word 'assumption' recommends itself as a better choice when saying what a free logic is.[16]

A second misunderstanding about free logic concerns the empty world. Long ago Russell complained about the impure character of classical logical truths such as

(9)　If every existent is unique then there exists something unique,

and

(10)　There exist objects identical with themselves.

[15] Bas van Fraassen, 'Singular terms, truth-value gaps, and free logic', *The Journal of Philosophy*, 67 (1966), pp. 481–495.
[16] Some may feel a tension between this diatribe on the possible advantages of the word 'assumption' over the word 'presupposition' in the characterization of free logic, and the earlier mildly derisive remark about Hintikka's lament about the word 'free' in free logic. The feeling should be dismissed. The expression 'presupposition' has a constituency familiar with its use, a lot of whose members regularly have gotten the wrong idea when the word 'presupposition' has been used to say what free logic is. But this sort of thing very likely has not been caused with any frequency by the word 'free'. In particular, no philosophical logician seems ever to have confused free logic with free algebra. Creating irrelevant worries is not among the more laudable aims of philosophical method.

According to Russell these statements are false if there are no objects at all. But, he thought, whether there exist any objects at all, and how many there are, are matters of fact rather than matters of logic. So holding statements (9) and (10) to be logical truths seemed to Russell to compromise the idea of logical truth as not dependent on matters of fact, a conception he himself encouraged. Now what should be noticed is that the principle of *Specification*, an essential feature of free logics, can be rejected and the statements (9) and (10) nevertheless consistently adopted as logical truths. It suffices to imagine the world to contain as its sole existing object the early twentieth-century solipsist logician Mary Caulkins.[17] Notice, first, that the way the principle of *Specification* was disproved earlier still works; the statement 'Every existent object is such that there exists something identical with it' is true of the imagined world, but the statement 'There exists something identical with Heimdal' is not. Second, because of Mary Caulkins (9) and (10) do not fail. The matter can be summed up in this way. If the world were empty (hence empty of existents), then, of course, all singular terms would fail to have existential import; but if there exist things in the world, it still might be the case that not any of them would be specified by singular terms such as 'Heimdal'. A free logic recognizing the empty world is easy to develop. It is called a *universally free logic*.[18] So, the point is, one might legitimately believe in a theory of singular inference that allows singular terms referring to no existing object while nevertheless not sharing Russell's sentiments about the logical impurity of statements (9) and (10). There is more than one kind of existence assumption, and free logic essentially has to do only with one of them.

A third misunderstanding about free logic concerns its relation to the classical logic of predicates. Many people – perhaps most – have the impression that free logic is an *alternative* to classical predicate logic in the sense that though all the principles of free logic are classically acceptable, some classical principles are not acceptable to the free logician. Nevertheless, free logic need not be considered an alternative to classical predicate logic.

---

[17] She is alleged to be responsible for a remarkable self-indicting response to a friend who met her as she was about to board a ship for England. 'Where are you going. Mary?', he asked. 'To a Solipsist convention in London', she replied.

[18] See Robert Meyer and Karel Lambert, 'Universally free logic and standard quantification theory', *Journal of Symbolic Logic*, 33 (1968), pp. 8–26.

The issue concerns what classical predicate logic is. Suppose it is formulated as Quine does it in his book *Word and Object*.[19] There the primitive vocabulary contains no constant singular terms, though it does contain variable singular terms that always refer to existent objects. These variables are not merely stand-ins for constant singular terms in logical formulas such as

F(x).

So the traditional logical principles of *Specification* and *Particularization* can and do hold for free variables. The only non-logical constants in Quine's version are predicates, that is, words and phrases like 'suave' and 'great economic brain'. Seen in this way no free logician challenges the conventional logic of predicates. So if constant *singular terms* such as 'Heimdal' and 'Bush' are added to the vocabulary, the resulting logic could be free, and an *extension* of classical predicate logic rather than an alternative.[20] Failure to perceive this sort of treatment of free logic, exemplified, for example, in the book by van Fraassen and me entitled *Derivation and Counterexample*,[21] flawed Susan Haack's discussion of free logic as an alternative to classical predicate logic.[22]

This does not mean that free logic can't be treated as an alternative to classical predicate logic. Imagine the free variables of classical predicate logic to function in part as stand-ins for singular term constants. If in the corresponding free logic, the free variables sometimes name no existent object,

∃x(x = y)

will fail for some variable singular term **y** and, hence, *Specification* also. Then the free logic would be an alternative to classical predicate logic thus conceived.[23]

---

[19] Willard Van Orman Quine, Word *and Object*, Wiley, New York (1960).

[20] This sort of treatment is quite analogous to those formulations of modal statement logic that simply add principles and rules for the expression 'it is necessary that' to conventional statement logic.

[21] Karel Lambert and Bas van Fraassen, *Derivation and Counterexample*, Dickenson, Encino, CA (1972).

[22] Susan Haack, *Deviant Logic*, Cambridge University Press, Cambridge (1974).

[23] This was indeed the treatment in my 'Existential import revisited', *Notre Dame Journal of Formal Logic*, 13 (1963), pp. 51–9.

It is often useful to think of free logic as an extension of classical predicate logic because it helps to isolate questions of singular inference. This strategy provides a clear view of the distinctive differences between free logic and the Frege and Russell theories of singular inference, and to different natural treatments of partial function names as kinds of singular terms without existential import.

The next misunderstanding about free logic concerns a confusion of formal syntax and (informal) semantics. Consider the purely quantificational part of Kit Fine's formal system of modal logic in his book *Modal Logic*.[24] That part includes, in addition to the postulate that all tautologous formulas are axioms, essentially the following axiom schemata:

MA1 $\forall x A(x) \supset (E!(t) \supset A(t))$, where t is a variable or a constant (with the usual restrictions when t is a variable);

MA2 $\forall x E!(x)$;

MA3 $\forall x(A \supset B) \equiv (\forall x \forall \supset \forall x B)$;

MA4 $A \supset \forall x A$, provided that if x occurs in A, it is bound;

and the rules

R1 From $A, A \supset B$ infer B;

R2 From A infer $\forall x A$

This formal syntax, except for stylistic variations and minor omissions, is essentially the same as the formal systematization of free logic set out in the paper 'Universally free logic and standard quantification theory' by Robert Meyer and me. Moreover, these schemata are read in the same way in both presentations. For example, **MA2** in both systems reads 'Every existent (or actual) object exists (is actual)'. But the quantificational logic thus syntactically represented in Fine's study is not free despite its near syntactic identity with the free logic represented in the paper by Meyer and me. How so?

Consider again an instance of the principle of *Specification*, say

---

[24] Kit Fine, *Modal Logic*, Blackwell (forthcoming). (I do not know whether the book in this reference of 1980 ever was published. But that is not important to the point being made here, since Fine's treatment in his book *vis-à-vis* the quantificational fragment is similar to the treatment in many current presentations of modal logic.)

the statement,

(11)   If every existent object is such that there exists something with which it is identical, then there exists something with which Bush is identical.

Free logicians reject the specimen statement because in the description of the real world there are singular terms referring to no existent (actual) object which, when substituted for the name 'Bush', turn the specimen statement into a non-truth – the expression '1/0', for example. But this can't happen in Fine's treatment because there are no singular terms that do not refer to existents (actuals) in the real world. The expression '1/0' is not a genuine singular term for Fine. Rather, the specimen statement, though true for all replacements of genuine singular terms by genuine singular terms, is rejected by Fine because it is not necessarily true. If it were necessarily true, then it would be true that

(12)   If necessarily every existent object is such that there exists something identical with it, then necessarily there exists something that is identical with Bush.

This cannot be the case in Fine's treatment because, though in every possible world every existent object is such that there exists something identical with it, there exists a possible world in which Bush is not identical with any existent object therein. Here is another reason why rejection of *Specification* though a necessary condition of free logic is not a sufficient condition.

The upshot of these remarks is that the condition

E!(t)

in Fine's restricted version of the principle of *Specification* is there for a different reason than it is in the syntactically similar statement schemata postulated by Meyer and me. For us, the condition **E!(t)** – the formal rendition of 't exists (is actual)' – was needed to block the invalid inference
from

Every existent is such that there exists something the same as it

to

There exists something that is the same as 1/0.

But for Fine the condition is needed rather to block the invalid inference from

Necessarily every existent is such that there exists something the same as it

to

Necessarily there exists something the same as Bush.

The situation here raises deep questions about logical truth informally understood, or, in Carnap's words, about the *explicandum* of logical truth. Here, informally, logical truth is truth by virtue of form alone. Given a listing of the logical particles – for example, 'not', 'and', 'every', and so on – this conception apparently amounts to the preservation of truth under simultaneous replacements of non-logical particles by all other non-logical particles of that kind. But for Fine, given the understanding of the informal notion of logical truth lately stipulated along with his policy on singular terms, there can be logical truths – for example (11) – that are not necessarily true. So apparently Fine must either abandon the current informal understanding of logical truth as truth by form alone or give up the equally strong intuition that logical truths are necessary truths par *excellence*.

It is not news that the number of philosophical logicians challenging the traditional conviction that all logical truths are necessary has been increasing. There are yet others who believe the impending conflict between logical truth and necessary truth in a development like Fine's is virtually a *reductio ad absurdum* of the view that only pure denoters are genuine singular terms. I, for one, confess to puzzlement over why a singular term, if it can fail to refer to an existent in a possible but unreal world, cannot fail to refer to an existent in the real (hence possible) world. Why, like its general counterpart 'same as 1/0', can it not fail to have existential import?

Since the subject of informal (or applied) semantics has now been raised, this is the appropriate place to explode another common misunderstanding about free logic. Opinion is about equally divided among non-connoisseurs that free logicians are either all Meinongians or all Russellians. Here Meinongians are those who accept non-existents and Russellians as those who deny them.[25] The current misunderstanding is probably a product of the place where one first runs into a free logic. Suppose one's first exposure is to a development like that in Burge's essay 'Truth and singular terms',[26] with the lack of an outer-domain in the description of the models, and a denotation function defined on the singular terms that is partial. Such an exposure may lead one to jump to the conclusion that free logic is inspired by a Russellian ontic disposition coupled with a rejection of Russell's theory of logical form. On the other hand, if one's initial exposure is via a paper like Leblanc's and Thomason's 'Completeness theorems for presupposition free logics',[27] replete as that is with outer-domains and a denotation function defined on the singular terms that is total, one might jump to the conclusion that free logic is Meinongian inspired. Neither of these two extreme conclusions is warranted. For example, in the paper by Meyer and me cited earlier there is an explicit denial of Meinongianism despite the existence of outer-domains; our outer domains were conceived as a set of expressions, the nominalistic counterpart of second intensions. Free logic, thus, does not presume any particular ontological inclination.

Another misunderstanding is the belief that free logic is somehow committed to the doctrine that existence is a predicate. This belief stems no doubt from the fact that many versions of free logic employ a primitive or defined symbol **E** – more often **E!** – as the formal counterpart of the English word 'exists' or other natural language equivalents of the English word, for example, 'existiert' in German. Nevertheless, the belief is mistaken.

In the first place, if one understands the traditional doctrine to concern nonlinguistic *things*, it is not at all clear that the occurrence of **E!** in the language of free logic, *even as a primitive predicate*, commits one to the view that existence is a property of individuals. It is not at all clear that existence is a property of *anything*, be it a property

---

[25] I have been identified many times as a Meinongian. Though having doubts about the fairness of treatment accorded Meinong by the majority of twentieth-century analytic philosophers, and despite an admiration for his inventiveness and imagination, I am not, nor have I ever been, a member of the Meinongian Party.

[26] Op. cit., Tyler Burge, 'Truth and singular terms' (1974).

[27] Hugues Leblanc and Richmond Thomason, 'Completeness theorems for presupposition-free logics', *Fundamenta Mathematicae*, 2 (1968), pp. 125–164.

of properties, or of propositional functions, or what have you. The point here is not the Quinian one that the expression **E!** is a general term and since general terms don't *refer*, it doesn't refer to properties. Rather the point is that even granted that general terms *stand for* things (though not perhaps in the sense names stand for things) as well as having extensions, the primitive symbol **E!** need not, and may not in fact, stand for properties of anything. Leonard, in his foundational essay, 'The logic of existence', allows only one of the symbols **E!**, or its complement **E!** (doesn't exist) to stand for a property. But in a free logic *even in principle* it is no more necessary for a given general term to stand for something than it is for any given singular term to refer to an existent – or to anything at all for that matter.

In the second place, if the traditional doctrine that existence is not a predicate is understood rather as the doctrine that the word 'existence' does not properly belong to the logical category of predicates, free logic still does not require that existence be treated as a predicate. On the one hand, Russell who introduced the symbol **E!** contextually in his theory of definite descriptions – understood as short for 'exists' – certainly did not construe the symbol as a predicate. On the other hand, there are developments of free logic that contain no symbol at all for 'exists' nor even envision paraphrases into the formal object language of natural language sentences containing that word as a part. My system in 'Existential import revisited' is a case in point.[28] Nor does any of this mean the earlier informal characterization of free logic suffers; that characterization appeals to statements containing the word 'exists' without taking a position on the logical form of such statements or the category of logical grammar to which the word 'exists' belongs. Of course, none of this rules out the possibility that **E!** (or 'exists') can be explicitly introduced as an expression in the linguistic category of predicate in a free logic.

Finally, consider a slogan almost synonymous with free logic in the minds of many, namely the slogan that free logic is the logic of irreferential singular terms. The slogan is a misleading way of expressing what *is* true about free logic. The issue concerns the meaning of the word 'irreferential' – or its fellows, the words 'non-denoting', 'empty', 'vacuous', and so on. Most people use the expression 'irreferential'

---

[28] Op. cit., 'Existential import revisited' (1963).

as synonymous with 'does not refer to an existent object'. This usage would not be so confusing were it not for the penchant of many
free logicians, following Russell, to equate the objects with the existent objects. But there are free logicians of a distinctly Meinongian
inclination who, rejecting the equation of the objects with the existents, thus are left asserting that the word 'Heimdal' is irreferential. For the Meinongian 'referential' has a broader meaning, and
'Heimdal', though not having existential import, nevertheless does refer to a nonexistent object, namely, the nonexistent object whose birth
was nine times as virginal as Christ's. This confusing talk of words and
phrases that are referential in a broader (or weaker) sense but not referential in a narrower (or stronger sense) should simply be dropped,
and 'irreferential' and its synonyms be reserved to mean 'does not
refer to anything at all existent or nonexistent'. A free logician whose
ontic proclivities are Meinongian is simply mis-characterized as one
who admits irreferential singular terms in his logical language. But he,
and his Russellian counterpart, still have at hand the very useful traditional phrase 'does not have existential import' to express agreement
vis-à-vis singular terms such as 'Heimdal' and '1/0'. Not that there
aren't persons who argue for the plausibility of nonexistent objects
but who also deny that every singular term refers – Terence Parsons,
for example.[29] In free logics, then, there may be expressions – 'Vulcan'
or 'the man born simultaneously of nine jotun maidens' or '1/0', for
instance – that are singular terms (contra Russell), don't have existential import (contra Frege), and may (Meinong) or may not refer to
some variety of nonexistent object (Parsons).

2. WHY FREE LOGIC?

Free logic is a logic of *terms*, general and singular. But the discussion has
not yet fixed sufficiently on what is novel about free logic. What is novel
is not its treatment of general terms, but its treatment of singular terms.
It breaks sharply with the traditional approaches to singular inference
in modern logic, the approach stemming from Frege, on the one hand,
and from Russell, on the other hand. So the question arises, 'What

---

[29] Op. cit., Terence Parsons, *Nonexistent Objects* (1980). Parsons, in contradistinction to
Meinong, thinks that 'The round square' refers to a nonexistent object, but 'Vulcan'
(the putative planet) does not refer even to a nonexistent object.

exactly are the Fregean and Russellian doctrines of "singular terms", and what is the doctrine in free logic?' (All of these views are about expressions occurring in at least an actual or reformed philosophical fragment of natural language.)

Consider the four expressions

(13)   Leo Sachse,
(14)   The President of the U.S. in 1979,
(15)   My mother's favorite joke,
(16)   Heimdal,

and the schema

(17)   There exists something the same as **s**.

Frege's position was, in effect, that the expressions (13) through (16), all of them members of the class of expressions he called 'Namen', are singular terms, that each of the expressions (13) through (16) can be legitimately substituted into the variable **s** in the validating schema (17), and that the resulting instances, for example, the statement

(18)   There exists something the same as Heimdal,

are true. Frege, in contrast to Meinong, but in agreement with many free logicians, considered the term 'Heimdal' to be irreferential. He thought it both "dangerous"[30] and odious to permit such singular terms in a language designed for the formulation and analysis of problems in science and philosophy. So he arbitrarily assigned to

---

[30] Concerning the danger of bearerless singular terms, Frege said:

> It is customary in logic texts to warn against the ambiguity of expressions as a source of fallacies. I deem it at least as appropriate to issue a warning against proper names that have no nominata. The history of mathematics has many a tale to tell of errors which originated from this source. The demagogic misuse is close (perhaps closer) at hand as in the case of ambiguous expressions. "The will of the people" may serve as an example in this regard; for it is easily established that there is no generally accepted nominatum of that expression. Thus it is obviously not without importance to obstruct once for all the source of these errors, at least as regards their occurrence in science.

> (See G. Frege, 'On sense and nominatum', in H. Feigl and W. Sellars (eds.], *Readings in Philosophical Analysis*, Appleton-Century-Crofts, New York (1949), p. 96.) This passage shows, by the way, that Frege's method of the artificial reference was not motivated simply by mathematical concerns, as is so often alleged.

singular terms like 'Heimdal' and 'the man who was born simultaneously of none jotun maidens' a referent, a certain set in one version of the theory, but a chosen individual in another version of the theory. This course has the result that since the statement in (18) turns out true, the false statement 'Heimdal exists' cannot be paraphrased by it.

Russell's view was, in effect, that *none* of the expressions (13) through (16) are singular terms, hence are not legitimate substitution instances of 's' in the validating schema (17), and that, contra Frege, the statement (18) is false. Accordingly, on Russell's view, the false statement 'Heimdal exists', containing the grammatically proper (but logically improper) name 'Heimdal' (qua truncated definite description) is acceptable shorthand for (18).

Free logicians believe that each of the expressions (13) through (16) is a singular term, that each expression is a legitimate substitution for 's' in (17), but that the statement in (18) is false. Hence, like Russell, the statement 'Heimdal exists' is properly paraphraseable as (18). In contrast to Frege, however, whose position on the logical form of the statement in (18) the free logician shares, the free logician does not arbitrarily assign any existent object to expressions such as 'Heimdal'.

These different attitudes about how to treat expressions such as those in the list (13) through (16) can be summed up in this way. In the interest of maintaining the validating character of (17), Frege and Russell brush aside one or another of the appearances exhibited by the statement in (18). Frege brushes aside the appearance of falsehood, and Russell brushes aside the appearance that (18) has the logical form of the statement in (17). Free logicians, on the other hand, in the interests of saving the appearances, reject the conviction that (17) is validating, and regard (18) as false.

In view of the preceding discussion, one might think that dissatisfaction with the Fregean and Russellian theories of singular inference served as the primary source of motivation for free logic. Yes and no. Dissatisfaction can be direct or indirect. *Direct* dissatisfaction arises from concern over whether a given theory is vindicated by its aims and validated by the evidence for it. For example, Russell's theory of definite descriptions is thought by many to be vindicated by its aims – for instance, the elimination of a need for nonexistent objects – though not validated by its sources of support – for instance, Russell's "proof" in *Principia Mathematica* that definite descriptions are incomplete symbols. (See Chapter 1.) *Indirect* dissatisfaction arises typically from the independent construction of a theory for a given end that, upon further consideration, turns out to be incompatible with an already existing theory. For example, Thomason's construction of a theory of singular inference adequate to the demands of modal discourse in the mid 1960s turned out to conflict with both the Russellian and Fregean approaches because it rejects the validity of the inference form,

$$t = s$$
$$G(t)$$
$$\therefore G(s),$$

an inference form which is validating both for Fregeans and for Russellians.[31]

Much of the dissatisfaction free logicians have expressed for theories of the Fregean and Russellian kinds, however, has been indirect. Free logic arose from other sources for other reasons in a mood of Cartesian independence. Now and then some free logicians showed their disagreement with Russell – Leonard, Hintikka, and Schock notably – but no one during the formative years of free logic began with a direct critical analysis of the traditional approaches of Frege and Russell.

Motivations come in layers. The deepest ones, sporadic, half-conscious and vague, seldom justify beliefs. These motives – the primordial ones – even if not always good reasons for beliefs, nevertheless often mitigate them. For example, there is a primordial intuition that logic is a tool that the philosopher uses (or *should* use, given Langford's famous complaint) and ideally should be neutral with respect to the ontological, epistemological ethical, etc., truth just as the various mathematical tools available to the empirical scientist – calculus, statistics, algebra, etc. – are presumed neither to create nor to pre-determine the empirical facts. The tool of logic is used to help decide among the various opinions what the philosophical truth really is – or at least it should according to primordial intuition (and Langford). So if there are preconditions to logic that have the effect of settling what there is

---

[31] Richmond Thomason, 'Modal logic and metaphysics', in *The Logical Way of Doing Things* (ed. Karel Lambert), Yale University Press, New Haven (1969).

and what there is not, they ought to be eliminated because they corrupt the ideal of logic as a philosophical tool. As a former practicing scientist, I admit the grip of this primordial intuition on that mildly sentimental phosphorescence that inspires and bedevils each of our allotted wits. But it would be a mistake to think all free logicians fall in this group. So to understand the whys and wherefores of free logic it is better to look at particular motivations for free logic operating nearer the surface of conscious deliberation than to concentrate on the more primitive inhabitants of the philosophical flux.

One specific source of motivation for free logic lies in a certain theoretical schizophrenia in the conventional logic of terms. Consider (again) the following pair of statement forms

(19)   Every existent object **x** is such that **G(x)**

and

(20)   **G(s)**

They are part of the apparatus of the classical logic of predicates, or as it is often called, the (first order) logic of (absolute) quantifiers. In (19) **G( )** is what remains of a statement when at least one singular term is left out – for example, the general term 'broods' in 'Heimdal broods'. In (20) **s** is what is left when 'broods' is deleted from 'Heimdal' broods. In the classical logic of predicates, statements of the form in (19) imply statements of the form in (20); indeed, they are respectively the antecedent and consequent in an instance of *Specification*. But, as noted above (and at greater length in Chapter 2), if **G( )** is replaced by the expression 'there exists something identical with it', and **s** is replaced by the expression 'Heimdal', or even the expression 'the object at position P on which no external forces are acting', then the resulting statement pairs are false. So, contrary to the classical logic of predicates, statements of the form (19) apparently do not imply statements of the form (20).

The response of the classical logician, nevertheless, is to maintain the implications between instances of (19) and (20) by insisting that they presume that the expressions one can plug into **s** in (20) refer to existing objects. It is, according to some scholars,[32] the dominant

---

[32] Ralph Eaton, *General Logic*, Charles Scribner, New York (1959), p. 223.

attitude of the medieval logician toward general terms in the square of opposition, an attitude rejected by the contemporary logician, but now reincarnated and applied to singular terms. This is the theoretical schizophrenia alluded to earlier.

That the conventional logic of singular inference is not free of existence assumptions with respect to its singular terms has explicit historical foundation because it is essentially Frege's response to Punjer when he (Punjer) challenged the validity of the inference:

> Sachse is a man
> So, there exists a man.

In response to Punjer's objection that this inference is valid only if supplemented by the premise 'Sachse exists', Frege says

If 'Sachse exists' means 'The word 'Sachse' is not an empty sound, but designates something' then it is correct that the condition be fulfilled. However, this condition is not a new premise, but the *obvious precondition* of all our words. The rules of logic always assume that the words used are not empty, that sentences are expressions of judgments, that one is not merely playing with words.[33]

For the free logician the central question is why statements including singular terms should be treated so differently from statements including only general terms. Why should the assumption of existential import be rejected in the case of general terms but not in the case of singular terms? Don't consequences similar to those resulting from the adoption of the assumption that all general terms are true of at least one existing object in the traditional logic of general inference obtain in the dominant contemporary logic of singular inference? For example, isn't application of logic to arguments containing statements such as 'The object at position P on which no external forces are acting maintains a constant velocity' prima facie prohibited? Shouldn't the methods of logic apply to reasoning containing expressions one may not be sure refer to any existing objects – as in the case of those astronomers who used the name 'Vulcan' in their explanations and conjectures before Vulcan was discovered not to exist? Indeed, ought

---

[33] Gottlob Frege, *Nachgelassene Schriften*, Felix Meiner, Hamburg (1969), p. 97. (My translation and italics.) Frege speaks of words in general as fulfilling the condition of non-emptiness, but the context of the discussion with Punjer makes it clear that he is talking about singular terms alone.

not the methods of logic apply to reasoning containing expressions
that one knows in fact don't refer to any existing object – such as the
reasoning of physicists using the definite description 'the body at position P on which no external forces are acting'? And finally, isn't there
a violation of the intuitive distinction between arguments whose validity requires existence assumptions – see Punjer's example – and those
whose validity doesn't – for instance, as in the case of the inference
from the statements 'Lincoln was the great emancipator' and 'Lincoln
brooded' to the statement 'The great emancipator brooded'?

The conflict between the contemporary approach to general inference and the dominant contemporary approach to singular inference
has not had the impact one would expect, probably because of the
availability of alternative methods for dealing with singular terms having no existential import. For example, there is Frege's method of
the chosen individual, Carnap's method of the null entity and more
recently Church's method of individual concepts. And, of course,
there is the philosopher's personal favorite, Russell's policy.

Beginning with Frege, as noted earlier, he arbitrarily picked an
existent – for example, the number zero – as the common referent
of Namen (singular terms) not ordinarily supposed to referring to
any existent. Technically the difference between singular terms such
as 'Russell', on the one hand, and 'Heimdal' on the other, is easily
discerned; the former belongs to the class of all singular terms not
referring to the number zero, whereas the latter belongs to the comple-
mentary class. This artifice for accommodating singular terms, ordinarily understood as having no existential import, is perfectly consistent
as Russell himself acknowledged as early as 1905. But it has unpalatable
peculiarities nevertheless. In the first place, the singular terms '0' and
'the number immediately preceding the positive integer 1', ordinarily
thought to refer to existents (at least by Frege), now wind up in the
class of singular terms like 'Heimdal'. In the second place, Heimdal
turns out to be the number which when multiplied by any number
equals itself. Given Heimdal's remarkable ancestry this might not seem
so strange, but surely it strains credibility to learn that zero has a
mother! In the third place, genuine problems now arise about the formerly uncontroversial *non-existence* of Heimdal, or equivalently about
the formerly uncontroversial *existence* of zero; one or the other belief
must go. For since Heimdal is the same as 0, either Heimdal exists or

0 doesn't. The general question is this; what *is* the proper analysis of the statements of the form '**s** exists' where '**s**' is a singular term? And there are other equally well known problems which may be ignored here because Carnap's method, though clearly in the spirit of Frege's chosen object method, sidesteps most of the oddities just outlined.

Carnap postulates a new entity, *the null thing* among (but distinct from all) the existents; it, rather than zero, is the common bearer of all singular terms normally thought not to have existential import.[34] Note that the singular terms '0' and 'the number immediately preceding the positive integer 1' no longer get misplaced in the category containing the singular term 'Heimdal'. Nor need zero now worry about the welfare of mother, a responsibility presuming the existence of one's mother, just because Heimdal now is identical with the null object, not zero. Finally, a straightforward analysis of statements of the form

    s exists, where **s** is a singular term,

is forthcoming. Let **n** refer to the null entity. Then '**s** exists' simply is '**s** $\neq$ **n**', though the singular term 'Heimdal' could just as well do the job of the name **n**. Carnap's method of the null object apparently does permit unrestricted application of the classical logic of terms, both general and singular. But it succeeds at an awful cost. It is bad enough that one must now endure the null object, a mysterious thing that, in Pico's words, "is beyond belief smiting the soul with wonder." But it is intolerable in another way. According to the Frege, and perhaps most of his successors, it is a precondition of the rules of logic that its singular terms stand for existents. But since the null entity is not different from itself, and so does not exist, the precondition is violated after all. Perhaps even worse, the quantifier word 'some' has to be given an interpretation with no existential force; otherwise the classically valid inference from the statement

[34] Rudolf Carnap, *Meaning and Necessity*, University of Chicago Press, Chicago (1947), pp. 36–37. Strictly speaking, the null entity view being examined here is a common misrepresentation of Carnap's actual view about the referent of the name 'a*', the name of "the null thing" (as Ermanno Bencivenga has reminded me). A close reading of Carnap's treatment of the name 'a*' early in *Meaning and Necessity* shows his strategy to be quite like Frege's chosen object method, and so is open to the (not original) complaints raised above against Frege's theory.

n does not exist

to the statement

Something does not exist

will fail. But these objections are not applicable to Church's method of individual concepts.

In his essay, 'Outline of a Revised Formulation of the Logic of Sense and Denotation (Part II)',[35] Alonzo Church says:

> the simplest formalized language results if the semantics is such that denotationless names are avoided. And it is believed that cases in which a natural-language sentence that contains a denotationless name seems nevertheless to have a truth-value are always better explained as instances of some logical or semantical anomaly, often *ungerade* usage.

For example, he notes

> In Fregean terminology, the name 'Pegasus' has an *ungerade* occurrence in such English sentences as 'Pegasus exists' ... and because the *ungerade* usage is logically anomalous, it is to be eliminated in a formalized language.[36]

This does not mean that one can't accommodate natural language statements containing names like 'Pegasus' in the formal language. Statements such as 'Pegasus exists' get paraphrased, in effect, as

$$e_{oi_1} P_{i_1},$$ where P_{i_1} is a name of the Pegasus-concept,

and

$$e_{oi_1}$$

expresses the actual existence of something of type I as presented by a concept of it of type i_1 (for example, the actual existence of the individual Bush as presented by the individual concept of Bush). The formal sentence, Church says:

> expresses the (probably false) proposition that Pegasus has or had actual existence.[37]

---

[35] Alonzo Church, 'Outline of a Revised Formulation of the Logic of Sense and Denotation (Part II)', *Noûs*, 8 (1975), p. 154.
[36] Ibid., p. 143.
[37] Ibid., p. 143.

Church's "method", though purporting to accommodate natural language statements containing names without existential import, is at odds with the intuition that there are natural language arguments containing such names whose validity requires an existence condition and those whose validity does not.

Consider, for instance, Punjer's example. He urged that the argument

> Sachse is a man,
> So, there exists a man,

is valid only given the additional premise

> Sachse exists.

Frege rejected Punjer's claim that the logical schema

G(s)
∴ There exists an **x** such that **G(x)**

is not validating on the ground that the schema presumes that the singular term **s** designates an existent. Both the validity of the schema construed as a schema about individuals, and the presumption, are honored in Church's proposal. The Churchian response to the claim that the argument from the statement 'Heimdal does not exist' to the statement 'There exists something that doesn't exist' shows the schema to be invalidating essentially is this. In the statement 'Heimdal does not exist', the name 'Heimdal' has *ungerade* usage and hence refers to the Heimdal-concept. The natural language statement then reads 'The Heimdal-concept does not present anything' and is true. Similarly the *ungerade* inducing natural language context '... exists' affects the paraphrase of the natural language statement 'There exists something that doesn't exist'. The quantifier expression 'there exists...' in this sentence is interpreted to range over individual concepts rather than individuals. So the natural language statement would read 'There exists an individual concept that doesn't present anything' and would be true because there certainly are individual concepts that present no individuals, the Heimdal-concept, for instance. The upshot is that the alleged counter-example to the classical schema, when properly construed, evaporates.

The Churchian treatment of singular terms like 'Heimdal' is objectionable on specific and general grounds. A specific case is this. Consider the statement 'Sachse exists'. On Church's proposal this statement may be formally paraphrased as

$$\mathbf{e_{oi_1}} \, \mathbf{S_{i_1}} \, .$$

Though Church provides no reading for the expression $\mathbf{e_{oi_1}}$ he says that the statement including it expresses the proposition that Sachse (actually) exists, hence, the same proposition presumably expressed by the natural language statement 'Sachse exists'. But this is puzzling. Propositions are often said to be *about* what the subject expressions of the statements expressing those propositions refer to, if anything. But then the statement 'Sachse exists' seems clearly to be about the theologian and not about the concept of Sachse. Yet the formal paraphrase *is* about the concept of Sachse, thus promoting the belief that the propositions expressed by the two statements are different. Perhaps the reaction to this will be that it is simply a prejudice that the name 'Sachse' in 'Sachse exists' refers to an individual, a prejudice that deeper analysis brings to the surface. This reaction conflicts with informal intuition, a standard that Church himself does not hesitate to invoke when circumstances are favorable.[38]

More general reasons are these. First, Church says that one ought to regard contexts of the form

s exists

as *ungerade*, and the occurrence of **s** in that context "anomalous". But the examples that Frege and others cite when speaking of such a context are relational contexts, like '__ believes that . . .', and so on. Singular existence contexts bear little logical resemblance to these, however. So an independent informal foundation for treating them as such is lacking. Second, Church's proposal is not so much a definite method as it is a program for a method; in key situations what is provided are hints instead of definite rules of paraphrase. One does not even know whether agreement in truth-value between natural language

---

[38] See, for example, his appeal to 'informal intuition' as partial support for his assertion that 'we should more properly say that for every individual concept, there are possible worlds in which it is vacuous', whenever confronted by the colloquial assertion that no individual necessarily exists. Ibid., p. 148.

statements and their formal paraphrase is always to be sought. Witness, for example, the difference between the natural language statement

(21)   Heimdal exists

and the Churchian formal paraphrase

(22) $\mathbf{e}_{oi_1} \mathbf{H}_{i_1}$

which have the same truth-value – that is, false – and the natural language statement

(23)   There exist individuals that do not exist

and its presumed formal paraphrase

(24) $(\exists \mathbf{x}_{i_1})(\sim \mathbf{e}_{oi_1} \mathbf{x}_{i_1})$

that seemingly disagree in truth-value, the former presumably being false, and the latter true; for it is not wholly unexpected news that there exist individual concepts that are (actually) vacuous.[39]

Church, though not denying the value of devising a logic with "denotationless" names,[40] defends the method of individual concepts on the ground of simplicity. This was asserted in the first sentence of the first passage quoted earlier from Church's essay. Surely it depends on what one counts important whether Church's approach is really simpler. For instance, if avoidance of intensions is important, then Church himself admits that Quine's treatment of singular existence statements is simpler because it requires fewer intensions. Thus, Quine would convert the name 'Vulcan' contained in the statement 'Vulcan exists' into the general term 'vulcanizer' and then treat the resulting statement 'Vulcanizers exist' as a general existence claim about individuals, that is, as 'There exist vulcanizers'. To be sure, admitting "denotationless" names into the logic complicates some of the rules but it does not bloat the ontology. Church's approach, however, inflates the underlying semantics with individual concepts even for a classical first order language with identity. In short, though Church's

---

[39] Ibid., p. 148. Church does not use the word 'paraphrase' but instead speaks of 'correcting the form'. This suggests that even statements that come close to contradiction, on their informal reading, might be turned into truths when corrected in form.

[40] Ibid., p. 154, fn 10.

method may be simpler for his purposes, it is surely not where the purposes of the free logician dominate, a fundamental purpose being continuity with the traditional treatment of general terms. There is no absolute measure of simplicity. The argument from simplicity lacks any real force in deciding between Church's proposal and that of the free logician.

There is another approach to the logic of statements containing expressions that do not refer to existents that has helped to blur the disparity between the contemporary theory of general inference and the dominant attitude toward singular inference. This approach, due to Russell, is the most popular among analytic philosophers of the twentieth century, and reflects best the dominant contemporary attitude toward singular inference. For Russell certainly acknowledges the precondition expressed by Frege in his discussion with Punjer, namely that the rules of logic assume that every *genuine* singular term refers to an existent thing.

Russell's approach to statements such as (13)–(15) and (18) is that they are only apparently logical subject-predicate statements. This is a respect in which his approach resembles the standard treatment of statements such as

(25)   Men are mortal,

a statement the tradition regarded as having the logical form of a subject-predicate statement, but which the conventional wisdom treats as a universal conditional, as paraphraseable into

(26)   Every existent thing is such that if it is a man then it is mortal.

On the other hand, Russell's reason for denying that the four statements above have the logical form of a (singular) subject-predicate statement is quite different from the motivation underlying the conventional rejection of the traditionalist's conviction that (25) is logically a subject-predicate statement.

Russell holds that the four statements above do not have the logical form of logical (singular) subject-predicate statements because none of the grammatical subjects stand for anything; contrary to appearance, they are not logically proper names, as is, perhaps, the word 'This' is in 'This runs'. They are, he says, truncated expressions of the form 'the so and so' (definite descriptions) and thus are "incomplete symbols". But in *Principia Mathematica* Russell maintains to all intents and purposes that expressions of the form 'the so and so' are not singular terms. So what he rejects is the assumption accepted in all other treatments of singular inference, namely, that the grammatical subjects in the four specimen statements are singular terms. Note that the conventional rejection of the traditional position that statements such as (25) are logically subject-predicate in form is not so based; the word 'men' survives analysis in the contemporary theory of logical form as a genuine general term.

Russell's approach dissolves the threat to the logical principle of *Specification*, and the principle in (17) lurking in the false colloquial statement (18), by denying that the word 'Heimdal' is a genuine singular term, and hence that (18) has the appropriate form of a genuine counterexample. In Chapter 1, the essential details of Russell's alternative theory of paraphrase were given, and eyebrows raised there about an approach that, in Hugues Leblanc's words, is "better on description than on acquaintance."

There is, however, an approach to the current problem, often confused with Russell's. It is due to Quine and deserves consideration because he, in contrast to Russell, holds that expressions such as 'Heimdal', though genuine singular terms, are a logical nuisance. He believes their introduction into what he calls the "canonical language" results in various kinds of inelegance, and so, in the interests of theoretical smoothness and simplicity, ought to be disallowed.[41]

The canonical language purports to be an idiom sufficient to the legitimate needs of science. The smoothness of that idiom would be affected, for example, by the admission of statements such as

(27)   Heimdal sings

which being neither true nor false, Quine believes, introduces truth-value gaps into the canonical language. But adoption of another truth-value, beside truth and falsity, to fill the gaps would complicate unnecessarily the two-valued canonical idiom. The truth-value gaps are caused by singular terms without existential import such as 'Heimdal', and, Quine notes, would not arise if two-valued canonical paraphrases

[41] Op. cit., W. V. Quine, *Word and Object* (1960).

for natural language statements containing them were available. Enter Russell. His technique of definite description provides the needed paraphrases, and indeed they are paraphrases in which singular terms such as 'Heimdal' do not survive.

A major complaint against Quine's approach is that it sacrifices the truth for convenience. If, prior to regimentation, 'Heimdal' is counted as a genuine singular term, then, by the Bolzano-Quine standard of logical truth, *it is true* that the statement

(29)   There exists something the same as Bush

is not logically true. So one would expect this fact about logic, surely an important ingredient in the conceptual scheme of science, to be reflected in the canonical language. Yet because singular terms have been banished from the canonical language for reasons of convenience, neither the non-logical truth of the statement nor the reason for judging it so can be represented in the canonical idiom. This dialectic raises the question whether the canonical idiom presented by Quine really does succeed in one of its major goals, namely clarifying the conceptual scheme of science.

Quine would, no doubt, be splendidly unmoved by this complaint, probably insisting that a truly global view of the matter would see no great conflict between truth and theoretical convenience. If so, one may wonder why the merging of truth and theoretical convenience doesn't support the view that the expression 'Heimdal' is more conveniently regarded as no singular term after all, an option that would indeed reduce Quine's view to Russell's view.

A final source of motivation for free logic to be considered concerns the foundations of epistemic, temporal, modal discourse and the like. Historically this sort of concern was right at the forefront of the development of free logic, as is evident in Jaakko Hintikka's pioneering treatise *Knowledge and Belief* [42] and in the treatment of individual constants (names) contemplated in Saul Kripke's influential essay 'Semantical considerations on modal logic'. [43]

---

[42] Jaakko Hintikka, *Knowledge and Belief,* Cornell University Press, Ithaca, NY (1962).
[43] Saul Kripke, 'Semantical considerations on modal logic', *Acta Philosophical Fennica,* 16 (1963), pp. 83–94.

Consider the most popular foundation for these various sorts of discourse – the possible worlds account – the inspiration for which is Leibniz in general conception if not in specific detail.[44] This kind of foundation provides a straightforward account of metaphysically necessary truth as truth in *all* possible worlds, and metaphysically possible truth as truth in *some* possible worlds. If the possible worlds are thought of as temporal stages, then temporally necessary truth amounts to truth in all temporally possible stages. If one supposes that the objects existing in one possible world may not exist in another world, it seems proper to explicate the statement 'Bush really exists' by the statement 'Bush exists in the real world', and to explicate the statement 'Pegasus might have existed but doesn't really' by the statement, 'Pegasus exists in some possible world, but not in the real world'. This style of explication has implications for statements about the reference of singular terms. For example, it suggests that the singular term 'Bush' refers to an existent in the real world, but the singular term 'Heimdal' does not. The upshot of these views about the reference of the singular terms 'Bush' and 'Heimdal' is, of course, that the statement

(30)   There exists something the same as Heimdal

is false in the real world, and hence that the conditional

(31)   If everything is such there exists something the same as it then there exists something that is the same as Heimdal

is false in the real world. The conditional is thus (and again) a counterexample to the classical principle of *Specification*. And the

---

[44] The phrase 'in general conception if not in specific detail' means that it is conceivable one might explicate necessary truth as truth in all possible worlds – in the general spirit of Leibniz's account of necessity, but adopt a view of possible worlds differing from Leibniz's (or a Leibniz-like) view. For example, Leibniz apparently holds that it is not true that a given possible object can occur in more than one metaphysically possible world let alone in all of them. This conviction is incompatible with the conception of the set of possible worlds as consisting of the same objects in each but differing only with respect to their configuration – a kind of popcorn machine picture of metaphysical atomism. (See R. Thomason, 'Modal logic and metaphysics' in *The Logical Way of Doing Things* (ed. Karel Lambert), Yale University Press, New Haven (1969), p. 127).

opinion here is reinforced by consideration of names like 'Heimdal' and definite descriptions such as 'the spherical non-spheroid'. Taken as singular terms, they refer to no existent in *any* possible world. The moral is that the quantificational part of the correct modal, or temporal, or epistemic logic is a free logic.

As noted earlier, the argument that the conditional above fails is resisted by some on the ground, reminiscent of, but not identical with, Russell, that expressions like 'Heimdal' and 'the spherical non-spheroid' aren't really singular terms, though the expression 'Bush' is. The expression 'Bush' is a singular term and not only refers to an object that exists in the real world but to that very same object in all possible worlds *even though that object may not exist in some merely possible world*. If the expression 'Bush' can fail to refer to an existent in some possible world other than the real world yet still be a singular term, it seems quite arbitrary to label 'Heimdal' a non-term simply because the real world is among the possible worlds in which its bearer doesn't exist.

The historically important sources of motivations discussed above, certainly do not exhaust the possible or actual or possible sources of motivation for free logic. Some, for example, find a basis for free logic in the theory of attribution,[45] some believe the correct foundation for mathematical intuitionism requires a free logic,[46] and yet others see it as a correct foundation for a theory of partial functions and the theory of definition.[47]

3. THE IMPORTANCE OF FREE LOGIC

Hintikka's 1970 essay 'Existential presuppositions and uniqueness presuppositions' contained an interesting opinion about the future of free

---

[45] Ronald Scales, *Attribution and Existence*, Ph.D. thesis, University of California, Irvine. University of Michigan Microfims (1969).

[46] Carl Posy, 'Free IPC is a natural logic: strong completeness for some intuitionistic free logics', *Topoi*, 1 (1982), pp. 30–43.

[47] See the last chapter of Karel Lambert and Bas van Fraassen, *Derivation and Counterexample*, Dickenson. Encino, California, 1972, the papers by William Farmer, Solomon Ferferman, and David Parnas in *Erkenntnis*, 43 (1959). pp. 279–338, and the paper by Raymond Gumb and Karel Lambert, 'Definitions in nonstrict positive free logic', *Modern Logic*, 7 (1997), pp. 25–56 (and corrections of printing errors in *Stanford University: CSLI Technical Report*, pp. 96–199).

logic.[48] Paraphrased from the original English, it reads:

I have some doubts as to how interesting an enterprise the study of logics free of existence assumptions with respect to their terms will turn out in the long run. It seems to me that a study of the role of existence assumptions in first-order logic will soon exhaust the general theoretical interest that there is in the area if conducted as a purely syntactical or semantical enterprise. It is true that there is a considerable variety of ways in which a semantics can be built for a 'free' first-order logic. However, a philosophically satisfactory comparison between them will in my opinion have to turn on a deeper conceptual analysis of the situation than the standard syntactical and semantical methods afford.

The emphasis in this passage on the need for philosophical investigations into the standard semantical foundations for free logic is well taken. Hintikka's observation, as a matter if fact, has proved quite prophetic. Here are some concrete examples.

Earlier various kinds of free logic were distinguished according to their treatments of simple statements containing singular terms without existential import, statements such as

(32) $1/0 = 1/0,$

for example. One of the major purposes in devising a formal semantics for a language is to explain the conditions under which a sentence of that language is true or false, if either. So which semantical basis for free logics emerges as the best kind of account, however the word 'best' is to be made precise, needs concentrated philosophical examination. Is it the one for positive free logics (semantics evaluating statements such as (32) true), or the semantical position of negative free logics (semantics evaluating statements such as (32) false), or, finally, the semantical foundation of nonvalent free logics (semantics evaluating statements such as (32) as having some other truth-value or none at all)? Examples of philosophical debate over the best semantical foundation are easily found in the literature. Here are two.

First, there is the debate between Tyler Burge, an exponent of negative free logic,[49] and Richard Grandy, an advocate of positive free

---

[48] Op. cit., Jaakko Hintikka, 'Existential presuppositions and uniqueness presuppositions' in *Philosophical Problems in Logic*, (ed. Karel Lambert), Reidel, Dordrecht (1970), p. 20.
[49] The debate can be followed in the papers of Grandy and Burge in the following order: first, Richard Grandy, 'A definition of truth for theories with intensional definite

logic, both of whom appeal to the theory of truth in support of their semantical accounts. A key point of contention between them concerns the truth-value to be assigned to statements of self-identity such as

(33)　　The man simultaneously born of nine sibling jotun maidens is identical with the man simultaneously born of nine sibling Jotun maidens.

Grandy holds (33) to be true, a verdict he believes to be required by the proper philosophical explication of truth developed in the Tarski style. Burge, however, presents an alternative explication of truth, again in the Tarski style, and shows that it supports the verdict that (33) instead is false. In response, Grandy has judged the issue between Burge and himself to lie even more fundamentally in distinct and philosophically deeper but unanalyzed intuitions about predication.[50]

This bare outline of the debate between Burge and Grandy suffices to establish it as an example of the kind of enterprise that surely falls under Hintikka's recommendation about what is desirable and important to the future of free logic.

A second example supporting Hintikka's point concerns a principle accepted by negative free logicians, but rejected by positive and

---

　　description operators', *Journal of Philosophical Logic*, 1 (1972), pp. 137–155; second, op. cit., Tyler Burge, 'Truth and singular terms' (1975).

[50] Richard Grandy, 'Predication and singular terms', *Noûs*, 11 (1977), pp. 163–167. Other grounds can be assembled in favor of Grandy's preference for positive free logic. For example, if identity is explicated in the usual way as property (or set) indiscernibility, then it is true that $t = t$ just in case it is true that for all properties P if $P(t)$ then $P(t)$, and in particular where t is the singular term 'the man simultaneously born of nine sibling Jotun maidens'. And from this the truth of (32) follows directly. Moreover, introduction of a singular term by definition, whose status *vis-à-vis* having existential import is not yet known, would seem to require a positive free logic. Otherwise not even the identity between the definiens and definiendum could be established. In negative free logics, it would seem, all definitions would be conditional on the the singular terms being defined having existential import, a severe restriction on those astronomers drawing inferences about the posited planet Vulcan prior to Leverrier's discovery of its nonexistence. Perhaps the upshot is that the sense attaching to the phrase 'is identical with' in Burge's treatment cannot be the 'usual' sense of that phrase. As a matter of fact, Lesniewski and his contemporary disciple Lejewski have long distinguished between a stronger and weaker use of the phrase 'is identical with', the weaker sense making (32) true and the stronger sense making it false. If they are right then the dispute between Grandy and Burge is only verbal, at least as far as statement (32) is concerned. I confess, however, that despite the inter-translatability of the alleged weak and strong senses of the phrase 'is identical with', it has never been clear to me why the strong sense is not merely mislabeled as 'is identical with'.

nonvalent free logicians. The principle is that a simple statement implies that its constituent singular terms always refer to existents – following the spirit of Russell if not the letter.[51] This principle might be called the *interdependence of being and being so.* It is rejected by Meinong. Indeed, its unacceptability is a direct consequence of his famous injunction that *being so is independent of being.* In the last part of the preceding century, many, for example, Terence Parsons, Richard Sylvan, and William Rapaport – have sided with Meinong in this famous philosophical debate. This is yet another example of the kind of philosophical enterprise responding to Hintikka's call for a philosophical analysis of the semantical foundations of free logic. If Meinong's view is correct, it would, of course, be a philosophical argument against the acceptability of negative free logics.

There is another respect, however, in which Hintikka's 1970 assessment is incorrect. Recall his pessimism about the yield of fruitful theoretical discoveries via standard syntactical and semantical procedures in free logic. Here his sense of things has not been so acute. A few modest examples will suffice to make the point.

Raymond Gumb in a paper published in 1979, and more recently and independently, Robin Dwyer in a Ph.D. thesis done at the University of California, Irvine, utilizing methods different from Gumb, have shown that Beth's definability theorem can be extended to free logic.[52] The subject of definability also provides a second example of a fruitful theoretical discovery in free logic via standard procedures. The basis of this discovery concerns *Specification.* It is deductively equivalent to

SB $(\forall x)(A \equiv B) \supset (A(s/x) \equiv B(s/x))$, with the usual restrictions if **s** is a variable.

---

[51] Russell and negative free logicians agree on the truth-value of statements like 'Heimdal broods', but disagree on the logical form of such statements. Russellians deny that 'Heimdal' is a singular term and hence that the statement 'Heimdal broods' is a predication, but negative free logicians regard the expression 'Heimdal'as a singular term and regard the statement in question as a genuine predication. Both hold the example statement to be false.

[52] Raymond Gumb, 'An extended joint consistency theorem for free logic', *Notre Dame Journal of Free Logic*, 20 (1979). Robin Dwyer, *Denoting and Defining; A Study in Free Logic*, Ph.D. thesis, University of California, Irvine (1988). Gumb's paper is especially important in another regard because he shows not only that many theorems of standard first order logic extend to free logic with classical identity, but also that some standard theorems do *not* extend to free logic.

It is easy to show that **SB** fails when singular terms like 'Heimdal', '1/0', 'Vulcan', and so on are present. For the statements

(34) $\forall x(x = x \equiv (x \text{ exists } \& x = x))$

and

(35) $\forall x(x = x \equiv (x \text{ exists } \supset x = x))$

are both true in free logic and standard predicate logic. Now suppose 'Heimal = Heimdal' is true. Then (34) yields, by **SB**,

(36)   Heimdal = Heimdal $\equiv$ (Heimdal exists & Heimdal = Heimdal),

which is false because Heimdal doesn't exist. On the other hand, if 'Heimdal = Heimdal' is false, then (35) yields, by **SB**,

(37)   Heimdal = Heimdal $\equiv$ (Heimdal exists $\supset$ Heimdal = Heimdal))
       which is true.

The instance of the *substitutivity of the biconditional* here is interesting for many reasons. Consider two of them. First, this principle is crucial in proofs that show that a predicate 'is the same as', defined in a vocabulary consisting of a *finite* set of primitive predicates, obeys the classical identity principles

Rl $s = s$
SI $s = t \supset (A(s) \equiv A'(t))$

where **s** and **t** are either variable or constant singular terms, and **A'** is like **A** except for containing **t** at one or more places where **A** contains **s**. This result has been cited by Quine as "a kind of justification of one's tendency to view '=', more than other general terms, as a logical constant"[53] and hence identity theory as a part of logic. A full account of the argument can be found in various places.[54] Now the important point is this: Since the *substitutivity of the biconditional* does not hold in free logic, the usual proof strategy doesn't work. The upshot is that conventional predicate logic creates an impression that turns out not to be so when the tacit existence assumption is made explicit. So there

---

[53] Op. cit., W. V. Quine, *Word and Object* (1960), p. 231.
[54] For example, W. V. Quine, *Set Theory and Its Logic*, Harvard University Press, Cambridge, MA (1969), p. 14.

is a question whether identity is definable in *any* language with only a finite set of primitive predicates. This in turn raises question about an important source of support for the Quinian philosophical belief that identity is a logical constant, and identity theory a part of logic.

In the same vein is the following observation. In free logic, the classical identity laws

(38) $\mathbf{A(s/x)} \equiv \exists x(x = s \,\&\, A)$

and

(39) $\mathbf{A(s/x)} \equiv \forall x(x = s \supset A)$

fail because **SP** fails. Therefore, the well-known device for eliminating singular terms from any given context, given a way of eliminating them from identity contexts, is not available in free logic. This consequence, and the preceding one, are less to be deplored as complications in the logical machinery than to be prized as helping to uncover the effect of existence assumptions in classical predicate logic. In general the question of which theorems and procedures of classical predicate logic carry over to free logic and which do not is important for precisely this reason: One learns from such an enterprise exactly where existence assumptions are important in classical predicate logic. The enterprise is of a piece with Quine's development and use of virtual classes to help determine which parts of set theory and arithmetic are free of ontic commitment to sets, that is, to help determine exactly where the existence of sets is crucial in set theory and its offspring. *Contra* Hintikka, the precise role of existence assumptions in classical predicate logic is still unknown, as the following consideration also helps to make clear.

Perhaps the most interesting feature of the apparent failure of the *substitutivity of the biconditional* in free logics concerns extensionality – at least of the substitution *salva veritate* sort. A long tradition in philosophical logic says that the extensions of predicates (general terms) are sets of n-tuples of *existents* the n-adic predicates (or general terms) are true of, if any at all. For example, the extension of the general term 'broods' is the set of all existent brooders, the brooders being what 'broods' is true of, and the extension of the general term 'unicorn' is the null set (because there are no existent unicorns).

A language that violates the rule that predicates having the same extension can replace each other in all statements without alteration of truth-value is non-extensional in the sense under consideration. The current instance of **SB** is a reflection in the object language of the extensionality condition just mentioned.

Extensionality is believed by some – Quine in particular and perhaps Frege also – to verge on a criterion of adequacy for a language that has, or could have, any pretensions at being an idiom sufficient to the needs of science. Moreover, it is an essential ingredient in Quine's (and many others') theory of predication and hence in Quine's theory of referential opacity. (See Chapter 6.) Now it is often claimed that conventional predicate logic with singular terms is extensional in the sense just explained. But free logic with the predicate **E!** is just standard predicate logic with its existence assumptions *qua* singular terms made explicit. So the suggestion that classical predicate logic with singular terms is extensional only because of suppressed existence assumptions is difficult to resist. True, the suggestion has force only if singular terms, constant or variable, are included in the logic of predicates. To be sure, avoidance of non-extensionality of the kind being discussed here can be secured for classical predicate logic by eliminating all constant singular terms and permitting variable singular terms never to occur unbound in logical sentences. (Unbound variables often operate as stand-ins for constant singular terms, and thus it is natural to think of them as not referring to existents only.) As noted earlier, it is indeed sometimes useful to think of predicate logic as thus restricted. But if the motive for confining predicate logic in the way just outlined is solely avoidance of non-extensionality, then the impression that non-extensionality has been secured rather artificially is hard to resist. Avoiding the problems created by singular terms via the sort of retrenchment outlined here is like championing withdrawal as *the* cure for schizophrenia.

It will be useful to look at some contributions and problems, old and new, technical and philosophical in free logic. Perhaps from such a litany the clearest perspective on the importance of free logic will emerge.

One of the major technical methods to come out of the development of free logic is the method of *supervaluations*. It was invented by

Bas van Fraassen in the mid 1960s.[55] Van Fraassen's method was designed in part to sustain the two convictions, first, that statements such as

(40) 5/0 is a number that broods or it is not the case that 5/0 is a

number, are logically true even though, second, the component

(41) 5/0 is a number

may have no truth-value. He accomplished this goal in the following way.

The manner in which the truth or falsity of simple statements is determined may be such that all of the statements of the language have a determinate truth-value, or only that some do. For example, one might say that a simple statement ('Jim Thorpe is an athlete') is true when its subject term ('JimThorpe') designates a physical object which is a member of the class of things associated with its predicate ('is an athlete'); that a simple statement ('Jim Thorpe is a Russian') is false if its subject term ('Jim Thorpe') designates a physical object which is not a member of the class associated with its predicate ('is a Russian'). If the language contains only singular terms that refer to existents, then all of the statements of the language will have a determined truth-value. On the other hand, suppose the language contains singular terms not associated with any existent object. Then, given the present characterizations of truth and falsity, some of the statements of the language – for example, the statement 'The Queen of the United States is an athlete' – would not have a determinate truth-value. This does not mean that they would not have a truth-value – a classical valuation requires that they have one – but only that their

---

[55] Bas C. van Fraassen, 'The completeness of free logic', *Zeitschrift für mathematische Logik and Grundlagen der Mathematik*, 12 (1966), pp. 219–234. The expression 'supervaluations' is not used in this paper, but was coined by van Fraassen later as more representative of the semantical development in the paper above. In the cited paper 'state models' was the term of art rather than 'supervaluations'. There were as a matter of fact unpublished completeness proofs in terms of inner domain-outer domain semantics. I had a Henkin style version in the early 1960s and I think Nuel Belnap also had one. Later this kind of semantics was independently exploited by Hugues Leblanc and Richmond Thomason, 'Completeness theorems for some presupposition-free logics', *Fundamenta Mathematicae*, 62 (1968), pp. 125–164.

particular truth-value would not be determined by the criterion. When we have a subclass of statements of a language whose truth-value is not determined according to some criterion, we have, in effect, a class of statements that are the classical surrogates of the truth-valueless statements of the user of that language. The intuitive idea here is that if a statement is regarded as truth-valueless, then if it were to be assigned a truth-value in a classical valuation, the assignment would be arbitrary; it could, with equal justification, be assigned truth or be assigned falsity.

This situation can be represented graphically as follows: suppose a language which consisting of only two atomic statements, **A** and **B**. Let **A** be the statement 'Jim Thorpe is an athlete'. It has the determined truth-value True. Let **B** be a truth-valueless candidate, say, the statement '5/0 is a number'. Let ∼ be the classical 'not' and **V** the classical 'or'. Then there are two possible classical valuations with respect to these two statements.

	C1	C2
A	T	T
∼**A**	F	F
B	T	F
∼**B**	F	T
A V ∼**A**	T	T
.	.	.
.	.	.
.	.	.
B V ∼**B**	T	T

Consider now the following (non-classical) valuation.

	S1
A	T
∼**A**	F
B	--
∼**B**	--
A V ∼**A**	T
.	
.	
.	
B V ∼**B**	T

S1 is an example of a *supervaluation*. It gives the truth-value assignments – truth, falsity or none at all – given the classical valuations. In general it is the product of a valuation rule obeying the principle that all and only those statements assigned **T**(rue) (or **F**(alse)) in all the classical valuations are to be assigned **T**(rue) (or **F**(alse)). The supervaluation **S1** represents what the classical valuations **C1** and **C2** have in common.

In the supervaluation **S1**, the statement **A** is assigned **T**, its negate is assigned **F**, **B** and its negate are assigned nothing, that is, they are truth-valueless, but **A V** ~**A**, and **B V** ~**B** are assigned **T**. The reason the compound **B V** ~**B** is assigned **T** is that no matter which value **B** were to get in the classical valuations, it always turns out true: Its truth-value is determined in the classical valuations. So it cannot be construed as truth-valueless, according to the intuitive idea that the truth-valueless statements are those whose truth-values are arbitrarily determined in classical valuations; supervaluations are a way of representing both the assignment of truth-values and the non-assignment of truth-values.

Customarily, 'logical truth' is defined as 'true in all classical valuations over all models'. In other words, no matter what domain of discourse and which associations between words and things is chosen, if a statement is logically true, it will be assigned True by any classical valuation. But in supervaluational semantics, to take of the cases where a compound my have truth valueless components 'logical truth' gets re-defined defined as 'true in all supervaluations over all models'. Indeed the logical truths in the logic of statements turn out to be the classical ones, but without the presumption that the atomic components actually have truth-values. For example, though **B V** ~**B** is logically true, it does not follow that **B** is true or **B** is false, as can be seen by looking at the corresponding classical valuations. It should be clear that no matter how the circumstances are varied, the same relationship between the classical valuations and supervaluations containing **B V** ~**B** will obtain. Hence **B v** ~**B** also is logically true in the redefined sense.

Van Fraassen's technique has received wide use. Not only has it been used in free logic, where it was the semantical basis of the first published proof of the completeness of positive free logic, but it has been used by van Fraassen himself to analyze the notion of presupposition introduced by Frege, utilized by Strawson in his attack on Russell's analysis of definite descriptions, and forms an essential part of Gordon Brittan's account of Kant's theory of Science.[56] It has received exploitation wherever there is reason to acknowledge truth-valueless statements and for whatever reason. Thus it has been used to reconstruct Reichenbach's quantum logic (see Chapter 9), to support Aristotle's conclusion that future tense statements such as 'There will be a sea fight tomorrow' are truth valueless though it is logically true that there will be a sea fight tomorrow or there will not, and in many other areas (and especially in the logic of vagueness.)[57]

An important feature of supervaluations, one that distinguishes it from most multi-valued logics, is the fact that they are not extensional, that is, co-valued statements do not always substitute *salva veritate*.[58] Moreover, there is a defect in van Fraassen's original approach to supervaluations evoking at least two semantical methods designed in part to overcome it. The defect is that van Fraassen's procedure is insensitive to the structure of simple statements containing singular terms that refer to no existent. Consider the statements

(42)   Vulcan is Vulcan

and

(43)   Vulcan rotates on its axis

many positive free logicians believe differ in truth-value, (42) being true, and (43) truth-valueless. Van Fraassen's original procedure treats them on a par. Without additional conditions on supervaluations, (42) and (43) are both arbitrarily assigned true or false in all classical valuations, and hence nothing in the corresponding supervaluations. But the complaint is that though this may be fair to (43) it is not to (42); that (42) should turn out true – indeed logically true – not truth-valueless.

---

[56] See Bas van Fraassen, 'Presuppositions. supervaluations and free Logic' in *The Logical Way of Doing Things* (ed. Karel Lambert), Yale, New Haven (1969), pp. 67–92, Gottlob Frege. 'Sense and reference', in *The Philosophical Writings of Gottlob Frege* (edited and translated by P. Geach and M. Black). Blackwell. Oxford, 3rd ed. 1980. pp. 56–78; P. F. Strawson, 'On referring', *Mind*, 59 (1950), pp. 320–344; and Gordon Brittan, *Kant's Theory of Science*, Princeton University Press. Princeton (1979).

[57] Kit Fine, 'Vagueness, truth and logic', *Synthese*, 30 (1975), pp. 265–300.

[58] See Ermanno Bencivenga, Karel Lambert and Bas van Fraassen, *Logic, Bivalence and Denotation*, Ridgeview, Atascadero, CA (1986).

Two ways of treating this problem are due to Robert Meyer and me, and later Ermanno Bencivenga.[59] The former incorporate the idea that some statements are true or false in virtue of facts about the world, some statements are true or false in virtue of facts about language, and some are indeterminate. Thus the semantics is developed so that the statement in (42) is true in virtue of the facts of language. The statement 'Pegasus is a cow' is false on the same grounds, but the statement in (43) gets no truth value in virtue of the world or in virtue of language, and hence is truth-valueless.

Bencivenga's strategy is different. He appeals to what he calls "mental experiments." The idea roughly is this: suppose a description of the world other than the actual world, but one in which Vulcan exists. Is it the case that the thing there specified is self-identical? Or that it rotates on its axis? If no matter what description of the world you look at, the answer is yes – as would be in the case of the first question – then the corresponding statement, (42), is assigned true. If for some worlds the answer is yes, and for others the answer is no – as it would be in the case of the second question – then the corresponding statement (43) gets no truth-value. The details of both of these approaches, inspired by van Fraassen's work, as well as the completeness proofs for free logic founded on them, are readily available in the literature.

Another interesting problem in free logic, a purely technical one. is the "Opium problem", so called because of the narcotic effect it has had on those who tried to solve it for over two decades. Leblanc and Meyer proposed an axiom set for free logic, based on my 1963 paper 'Existential import revisited' containing neither identity nor a primitive existence predicate. The axioms, closely enough, are all tautologies, and the prefacing of zero or more universal quantifiers to the following kinds of formulas:

A1 A, if A is a tautology.
A2 $A \supset \forall xA$, if x is bound in A
A3 $\forall x(A \supset B) \supset (\forall xA \supset \forall xB)$
A4 $\forall x(\forall yA \supset A(x/y))$

[59] Robert Meyer and Karel Lambert, 'Universally free logic and standard quantification theory', *Journal of Symbolic Logic*, 33 (1968), pp. 8–25, and Ermanno Bencivenga, 'Free semantics', *Boston Studies in the Philosophy of Science*, 47 (1980), pp. 219–229.

The theorems are all instances of **A1–A4** by re-lettering of bound variables and whatever is yielded by the inference rule *Modus Ponens*. Leblanc and Meyer ostensibly proved the semantic completeness of the above system,[60] but, it turned out, their proof relied on the assumption that the principle of universal quantifier commutation, that is,

UQC $\forall x \forall y A \supset \forall y \forall x A$

is provable in the Leblanc and Meyer. Attempts to derive **UQC** in the LebLanc and Meyer treatment are destined to fail because Kit Fine finally proved that **UQC** is independent of the Leblanc-Meyer axiom set (though it is not so in my 1963 treatment because that treatment had the classical axioms for identity).[61]

One reason **A1–A4** + **UQC** is important is because it is an essential ingredient in the proof that the predicate **E!**(xists) is not eliminable in the formulation of positive free logic, **FQ**, without identity by Meyer and Lambert.[62] **E!** is justifiably eliminable, if identity is available, via the definition,

DE!   **E!**(s) for $\exists x(x = s)$, where **s** is a singular term,

as Hintikka first showed. The question is whether there is any formula that can do the job of '**E!**' in those developments of free logic without identity but containing a primitive predicate **E!** such that all instances of

$\forall x(El(x))$

and

$(\forall x A$ & **E!**(s)$) \supset A(s/x)$, where **s** is a constant singular term,

are logically true but nevertheless are consistent with the sentence

$\sim$**E!**(s)

---

[60] Hugues Leblanc and Robert Meyer, 'On prefacing $\forall x A \supset A(y/x)$ with $\forall y$: A Free quantification theory without identity', *Zeitschrift für mathematische Logik und Grundlagen der Mathematik*, 16 (1970), pp. 447–462.

[61] Kit Fine, 'The permutation principle in quantificational logic', *Journal of Philosophical Logic*, 12 (1983), pp. 33–37.

[62] Op. cit., Robert Meyer and Karel Lambert, 'Universally free logic and standard quantification theory' (1968).

that is, consistent with the nonexistence of any purported existent as a referent of '**s**'. This is the question Meyer, Bencivenga and I answered in the affirmative.[63] Here is a sketch of the argument.

Consider a language **L** whose alphabet consists of variables, constant singular terms, predicates, the logical signs

$$\sim, \&, \forall, (,),$$

and whose formation rules and transformation rules are those of the Leblanc-Meyer system outlined above. Consider another language, **L + E**, whose alphabet is exactly like the language **L** except for containing in addition the monadic predicate **E!**, whose formation rules are exactly like those of **L**, but whose transformation rules are the classical tautologies **A1–A4, UQC,** *Modus Ponens* and the following axiom schemata

A5 $(\forall xA \ \& \ E!(s)) \supset A(s/x)$, where **s** is a constant singular term,
A6 $\forall x(E!(x))$.

The logic of the language **L + E** just outlined is (essentially) an expansive version of **FQ**. Now define a theory **T** as **FQ** plus the postulate

P1 $\sim E!(s)$ for every constant singular term *s*.

The proof of the ineliminability of '**E!**' in positive free logic, indeed in any free logic, depends on two lemmas. The first is that all sentences not containing the monadic predicate '**E!**' and deducible in **T** are deducible in **A1–A4 + UQC**, et. al. In other words

Lemma 1: The theory **T** is a conservative extension of **A1–A4 + UQC**, et. al.

The argument may be sketched as follows. Let **C** be a sentence of **T** not containing **E!** (hence expressible in the language **L**). Let $A_1, A_2,...,A_n$ (where $A_n =$ **C**) be a proof of **C** in **T**. That proof can transformed into a proof in **A1–A4 + UQC**, et. al. by replacing any occurrence of **E!** prefixed to a variable in any **A**, by the sentence $P(x) \lor \sim P(x)$ where **P** is a monadic predicate, and any occurrence of **E!** prefixed to a singular term constant in any A_i, by the sentence

[63] Robert Meyer, Ermanno Bencivenga and Karel Lambert, 'The ineliminabilty of E! in free quantfication theory without identity', *Journal of Philosophical Logic*, 11 (1982), pp. 229–231.

$P(s)$ & $\sim P(s)$ where P is a monadic predicate and s is a singular term constant. A simple induction suffices to establish the claim. (The only problematic case, perhaps, is an A_i, of the form

$$\forall y((\forall xA \ \& \ E!(y)) \supset \sim A(y/x)).$$

But then A_i gets transformed into $\forall y(\forall xA \ \& \ (P(x) \ V \sim P(x))) \supset A(y/x))$ which follows immediately from **A4** in **A1–A4** + **UQC** et. al.

The second lemma says

Lemma 2: There is no sentence of the form $B(x)$ of **L** such that all instances of both the schemata $\forall x(B(x))$ and $((\forall xA \ \& B(s/x)) \supset B(s/x)$ are provable in **A1–A4** + **UQC** et. al.

To prove this lemma, it must be shown, first, that no sentence of the form $\forall x(A \ \& \sim A)$ is a theorem of **FQ**. Call this claim, **CL1**, and note that it follows easily by semantical counterexample. For $\forall x(A \ \& \sim A)$ holds only in the empty universe, but, as Meyer and Lambert showed in 1968, if A is a theorem of **FQ** it holds in all universes of discourse including the empty one.[64] Next, for reductio, assume the negation of the lemma. Then for any monadic nonempty predicate P, $(\forall xP(x) \ \&$ $B(s/x)) \supset P(s)$ (where s is a singular term constant) is a theorem of **A1–A4** + **UQC** et. al. and hence of **FQ** also. So there will be a proof of the sentence in question in **FQ**. Substituting everywhere $E!$ for P in the proof in question one obtains immediately a proof also of $(\forall x(E!(x)) \ \&$ $B(s/x)) \supset E!(s)$ in **FQ**. But in virtue of **A6** in **FQ**, it follows that $B(s/x) \supset$ $E!(s)$, where s is a singular term constant. Since **FQ** is contained in theory **T** the last mentioned sentence is a theorem of **T**, and, thus, so is $\sim B(s/x)$, in virtue of **P1**. Since $\sim B(s/x)$ contains no occurrence of $E!$, it is a theorem of **A1–A4** + **UQC** et. al. by **Lemma 1**. Since s is arbitrary, then by Universal Generalization, a derivable rule in **A1–A4** plus **UQC** et al. $\forall x \sim B(x/s)$ is a theorem in therein. But, by assumption, $\forall x(B(x))$ is also a theorem therein, and hence also $\forall x(B(x) \ \& \sim B(x))$, and thus is theorem of **FQ**. But this contradicts **CL1**.

To show that $E!$ is not eliminable in **FQ**, suppose there is a sentence in the language **L** + **E** that does the job of $E!$ in **FQ**. Call it $B(s)$, where s is a singular term constant or variable, such that it contains no occurrence of $E!$. Then the results of replacing $E!(x)$ by $B(x)$ in **A6**, and

---

[64] Op cit., Robert Meyer and Karel Lambert, 'Universally free logic and standard quantification theory' (1968).

E!(s) by **B(s)** in **A5**, should also be theorems of **FQ**. Since **B(s)** contains no occurrence of **E!**, all instances of the replacements of **A5** and **A6** in question should be deducible in **A1–A4** + **UQC** et. al. (by **Lemma 1**). But this contradicts **Lemma 2**. So there is no sentence of the form **B(s)** not containing **E!** that does the job of **E!** in **FQ**, nor, therefore in the theory **T**. Note that the theory **T** reflects the conditions that were to be met by any successful surrogate for **E!**, if such there be. This ends the proof sketch.

The moral is that there is no way of asserting the existence of Bush, or the nonexistence of Heimdal, except by means of the predicate **E!** in a theory of singular existence based on **FQ** (but without identity, of course). Contrast this with the case of general existence as reflected in the statement 'Men exist'. 'Exist' in such contexts is eliminable; for example, 'Men exist' paraphrases as 'There are men' in both standard and free quantification theory without identity. This fact has played a part – though a suspect one – in the vexing philosophical problem whether existence is a "predicate."

Another important consequence of the ineliminability in free logic without identity is that it can be used to show that Quine's argument in classical predicate logic that identity can always be eliminated in a finite bank of predicate must fail in free logic. For suppose identity could be reduced a la Quine to a finite bank of predicates in free logic. Then **E!** could be eliminated via that same finite bank of predicates. But in virtue of the ineliminabilty theorem above, **E!** is not eliminable in any such bank of predicates, and hence neither is identity. This counts against Quine's view that one reason identity can count as a "logical" word is because it is always available implicitly in any finite bank of predicates.

No discussion of the importance of free logic can ignore the topic of definite descriptions. And in fact it is a subject matter to which much attention has been given in the literature. There exist many different formulations of free definite description theory already (see Chapter 5), and no doubt others lurk only just beneath the surface of publication. One reason free logic is important as a medium in which to formulate theories of definite descriptions is that only by means of it can one compare directly the different *logics* of definite descriptions of Russell and Frege without a detour through their respective philosophies of language. Important features of Russell's logic of definite

descriptions, its particular truth conditions for statements containing
definite descriptions and its doctrine of scope, can be reproduced in
free logic without having to drop definite descriptions from the category of singular terms. These essentials of Russell's logic of definite
descriptions have been developed in free logic independently by many,
the earliest being Rolf Schock, then later by Ronald Scales and Timothy
Smiley (among others). Similarly, the spirit of Frege's particular logic
of definite descriptions has also been developed in free logic by van
Fraassen and me, and (in effect) also by Scott, minus Frege's troubles
with contexts of the form **E!(s)**.[65] One consequence of the comparison is the not unsurprising fact that the underlying identity theory
must be non-classical in the Russellian theory, another is the discovery
that the Frege theory of definite descriptions can be so formulated in
free logic that a version of the principle of the reflexivity of identity
can be eliminated from the primitive foundations of classical identity
theory. Finally, the treatment of contexts of the form **E!(ix(A))** can be
rendered in the same way in either tradition in Russell's manner sans
Russell's belief that definite descriptions are not singular terms (see
Chapter 3).

By the early 1970s there existed already a wealth of positive free
description theories. Bas van Fraassen early speculated that they could
all be arranged in a one-dimensional hierarchy much in the way Lewis
set up his hierarchy of propositional modal logics. But a few years
ago Peter Woodruff and I noted that a system of positive free definite description theory published due to Richard Grandy violated van
Fraassen's picture of what he called 'the spectrum' of positive free description theories'. The hierarchy of positive free description theory is
at least two dimensional. Moreover, the most straightforward semantical account of the new picture involves recourse to existent and nonexistent objects, extensions and comprehensions, and a sharp distinction between the truth conditions for identities containing definite

[65] See op cit., Rolf Schock, *Logics Without Existence Assumptions*, 1968; op.cit., Ronald
Scales, *Attribution and Existence* (1969); Timothy Smiley, *Lectures in Mathematical Logic*,
Cambridge University Press, Cambridge (1970); op. cit., Karel Lambert and Bas van
Fraassen, 'On free description theory' *Zeitschrift für mathematische Logik u, Grundl. der
Math* (1967); and Dana Scott, 'Existence and description in formal logic', in *Bertrand
Russell: Philosopher of the Century* (ed. Ralph Schoenman), Allen and Unwin, London
(1967).

descriptions without existential import and those for predications of a similar sort (see Chapter 5).

There is a phase of free logic in need of serious technical investigation. To make the issue clear, recall a remark at the beginning of the previous section. There the point was made that what was distinctive about free logic was its position on singular terms, that what it had to say about general terms was quite conventional. That remark about general terms in free logic was unduly glib. To be sure the treatment of general terms in free logic is quite conventional but only because, strictly speaking, there was until recently no theory of general terms in classical predicate logic.[66] There were hints but no theory; and the same goes for free logic. So given that free logic is a theory of terms, half of the enterprise has not been adequately addressed let alone accomplished formally, syntactically or semantically. Until this is done, it will be hard to assess the claim that free logic in part is a logic free of existence assumptions with respect to its *general terms*, there being no formal account of general terms in free logic.

Consider the conventional symbolism of classical logic, say, the formula

F(t), where **t** is a singular term,

and its use in paraphrasing a statement from natural language. The question now is what part of the paraphrase does **F( )** represent? If the statement is one like

Nixon broods,

then **t** is the singular term 'Nixon' and **F( )** is the general term 'broods' thus encouraging the idea that **F( )** in **F(t)** stands for general terms just as the letters **S** and **P** do in the traditional schema

all **S** are **P**.

But the idea is unjustified when one examines examples such as

Nixon is a brooder.

---

[66] A deficiency recently eliminated by Robert Stalnaker in 'Complex Predicates', *The Monist*, 60 (1977), pp. 327–340.

Here **t**, again, is the singular term 'Nixon', but **F( )** now is the phrase 'is a brooder'. It not only is not a general term, it is not a term at all.[67] In the modern logical tradition beginning with Frege, the old 'is' and 'is a(n)' of copulation – the 'is' in 'Bush is limited' and 'is a' in 'Cheney is a conservative politician' – have been absorbed into what now is commonly called the *predicate*. In doing so the capacity to isolate those parts of the simple sentence which are *genuine* general terms both from those parts which are singular terms and from those parts which are neither was lost. But the need for an apparatus of general terms is quite pressing. Formal developments in some quarters of modal logic are perhaps expressions of that need as in the apparatus designed to distinguish between *de dicto* and *de re* occurrences of the modal operator 'it is necessary that'. Employing the cap notation from set theory, Stalnaker and Thomason[68] express *de dicto* occurrences thus

$$N(û(Fu)t)$$

where **N** is 'Necessarily', **F** is a predicate (as lately described), **u** is a variable and **t** is a singular term. So this sentence schema reads: necessarily **t** has the property of being (an) **F**. *De re* occurrences of necessity are expressed thus

$$û(N(Fu))t$$

which reads: **t** has the property of being necessarily (an) **F**. The distinction is needed to express the ambiguity in natural language statements such as

God necessarily exists

which could mean

$$N(û(E!u)g)$$

that is, 'Necessarily God has the property of existing', or could mean

$$û(NE!u)g$$

[67] See Henry Leonard's 'Essences, attributes and predicates', *Proceedings and Addresses of the American Philosophical Association*, vol. XXXVII (1964), pp. 25–51.

[68] Robert Stalnaker and Richmond Thomason, 'Modality and reference', *Noûs*, 2 (1968), pp. 359–372.

that is, 'God has the property of necessarily existing', two statements that may differ in truth-value.

There are all sorts of logical reasons for wanting a formally developed theory of general terms even at the level of elementary predicate logic. Two will suffice here, the second one of which will bear directly on the final topic, the problem of predication.

First, some philosophical logicians (Ronald Scales, for instance)[69] have felt the need of a theory of general terms to be able to express an ambiguity they think latent in natural language sentences such as

Heimdal doesn't exist

which could mean

It is not the case that Heimdal exists

or

Heimdal is a non-existent.

Using the Stalnaker-Thomason notation, the first might be expressed as the true

$\sim(\hat{u}(E!u)Heimdal)$,

and the second as the false

$\hat{u}(\sim E!u)Heimdal$

there being no Heimdal for the general term $\hat{u}(\sim E!u)$ – that is, the general term 'non-existent' – to be true of.[70] The situation is similar in some respects to Stalnaker-Thomason. In general, the abstraction principle

$\hat{u}(Fu)t \equiv Ft$

---

[69] Op.cit., Ronald Scales, *Attribution and Existence* (1969).

[70] One may wonder why, when the traditional square of opposition with its precondition that all general terms be true of at least one actual thing was rejected, alternative treatments of general terms acknowledging that precondition were not developed as in the case of singular terms. Surely part of the reason is that no apparatus in the new symbolism for general terms was available and hence questions like the Russellian question whether expressions such as 'the round square' are really singular terms could not profitably be asked about expressions such as 'round square'.

that is,

   t is such that it is (an)**F** if and only if **t** is (an)**F**,

is suspect where elementary predicate logic contains, or is supplemented by, singular terms having no existential import.

Finally, return to the topic of predication, a topic that arose earlier in the debate between Burge and Grandy on the correct semantical foundation for free logic. Recall Grandy's view that the issue between Burge and himself turns on a deep-seated difference about the nature of predication, but about predication as kind of logical form, not as a certain relation between words and things. That is the issue is not about predication in the sense in the general term 'pretty' is *predicated of* the girl next door in the statement 'The girl next door is pretty'. Rather it is about a predication as a construction in which a *general term* is joined to a singular term to form a sentence that is true or false according as the general term is true or false of the object specified by the singular term. The point about general terms then is this: The theory of predication so stated requires general terms, and to the extent that there is no formally clear theory of general terms there is no formally clear theory of predication. But such a theory is sorely needed by certain approaches to free logic.

Consider, for instance, the statement

Heimdal broods.

It counts as a predication qua logical form in any version of free logic. Now semantical versions of free logic generally fall into two kinds: first, those in which every singular term is assigned an object though, as in the case of the singular term 'Heimdal', the assigned object may be nonexistent. Second, there are those in which singular terms such as 'Heimdal' are not assigned *anything* existent or nonexistent. The concept of predication qua logical form expressed above in Quinian language,[71] is readily applicable to free logics of the first kind, because there is something – the nonexistent object Heimdal, for instance – for the general term 'broods' to be true (or false) of, hence a ready means

---

[71] It should be noted that though the language is Quine's, the theory of predication so expressed is not. Op. cit., W. V. Quine, *Word and Object* (1960), p. 96.

for telling when the statement 'Heimdal broods' is true (or false). But in the case where 'Heimdal' is regarded as standing for nothing at all, the truth-value of the statement 'Heimdal broods', if any at all, cannot depend upon some relation between general terms and entities, there being no such entities. So whereas the first semantical approach to free logic is confronted with the ontological problem of making sense out of nonexistent objects, a problem which Terence Parsons' book *Nonexistent Objects*[72] deals with directly and ingeniously, the second approach is confronted with the problem of saying exactly what predication is. Grandy is a representative of the first kind of approach to free logic, Burge of the second. It is now clear just how pivotal the notion of predication *is* in the debate between Grandy and Burge.

Hintikka notwithstanding, consequences, philosophical or otherwise, of making existence assumptions explicit in predicate logic continue to emerge. Indeed, it has often come as nothing short of a revelation that some principles and procedures of predicate logic depend crucially on existence assumptions. Such seems to be the case, for instance, in the recent discovery that the paradoxical consequences of a certain natural principle of definite descriptions basic to all free theories of definite descriptions are dependent on existence assumptions tied to singular terms in classical predicate logic. Moreover, avoidance of those paradoxical consequences can be used to explain the various traditional approaches to the logic of definite descriptions from Frege to Russell in essentially the same way that various approaches to set theory can be seen as different ways out of the paradoxical principle of naïve abstraction (see Chapters 2 and 3).

---

[72] Op. cit., Terence Parsons, *Nonexistent Objects* (1980).

9

Logical Truth and Microphysics

1. LOGICAL TRUTH AND TRUTH-VALUE GAPS

Elementary microphysical statements[1] can be neither true nor false without violating the classical codification of statement logic. The existence of such a possibility depends upon a revision in the standard explication of logical truth, a revision more harmonious with the idea of argument validity as *merely* truth-preserving. The revision in question, in turn, depends upon Bas van Fraassen's investigations into the semantical foundations of positive free logic,[2] a species of logical system whose philosophical significance was first made plain in Henry Leonard's pioneering study of 1956, 'The Logic of Existence'.[3]

Students who steadfastly refuse to accept an argument as valid unless all of the component statements are in fact true frustrate teachers of logic. "Now look!" the teacher may heatedly emphasize, "the validity of an argument has to do with its *form* alone. So to say that an argument

---

[1] The word 'statement' refers to a sentence without free variables of the sort found in the simple languages studied in elementary logic. The simple languages in question are first-order languages, languages that do not contain statements like 'There is a property had by everything' or 'Every property Jim has, the king has'. The first-order languages of concern here have a vocabulary consisting of variables, connectives, quantifiers, predicates, and both referential singular terms (e.g. 'Jim Thorpe') and singular terms without existential import (e.g. 'the Queen of the U.S.'). It will not be too misleading to think of them as certain restricted fragments of English.

[2] Bas C. van Fraassen, 'The completeness of free logic', *Zeitschrift fuer mathematische Logik und Grundlagen der Mathematik, 12* (1966), pp. 219–234.

[3] Henry S. Leonard, 'The logic of existence', *Philosophical Studies,* 7 (1956), 49–64.

is valid is to say *only* that if its premises *were* true its conclusion *would* also be true!" But, then, not only is the argument from the pair of false statements

Jim Thorpe was Russian

and

If Jim Thorpe was Russian, he was a Bolshevik

to the false statement

Jim Thorpe was a Bolshevik

valid, but so is the argument from the pair of (allegedly) truth-valueless statements

The Queen of the United States dreamed she was being led down a bridal path by a gorilla

and

If the Queen of the United States dreamed she was being led down a bridal path by a gorilla, she desires to marry a man named 'Harry'

to the (allegedly) truth-valueless statement

The Queen of the United States desires to marry a man named 'Harry'.[4]

The characterization of validity as merely truth-preserving does not imply that the constituent statements of an argument are true *nor even that they have any truth-values at all* despite one's beliefs about whether there are or are not any truth-valueless statements. Why this should be embarrassing, given the dominant account of logical truth, is easily explained.

Classically, logical truth is taken to be a measure of validity; an argument is valid if and only if the conditional having the conjunction of the premises of the argument as antecedent and the conclusion of the argument as consequent is logically true. But, even when interest is restricted to the classical logic of statements, the prevailing version of

---

[4] Wesley Salmon bears ultimate responsibility for this example.

logical truth excludes truth-valueless statements. This is so even though the classical statement of logical truth – truth under all assignments of truth-values to the atomic components – does not, on the face of it, exclude the possibility that *no* assignments to the atomic components are made, that the atomic components have no truth-values!

Classical two-valued logic recognizes statements other than conditionals as logically true. For example, any statement of the form

(1)   **A** or it is not the case that **A**

is logically true. The prevailing explication of logical truth thus is framed to include statements like 'Jim Thorpe is an athlete or he is not' as logically true. However, given the Tarski adequacy condition for truth, instances of (1) including the truth-valueless 'The Queen of the United States dreams' as a component must be excluded. For suppose otherwise; assume, for an arbitrary statement **S**, that

(2)   It is logically true that **S** or it is not the case that **S** is logically true, but it is not the case that **S** has a truth-value

If the statement named by

   S or it is not the case that **S**

is logically true, it is true. By the Tarski adequacy condition, therefore,

(3)   **S** or it is not the case that **S**.

Moreover, the Tarski adequacy condition, and the convention that

   A is false *means* it is true that (it is not the case that **A**)

yields a conditional of the form

(4)   If **S** or it is not the case that **S**, then is the case that **S** is true or it is the case that **S** is false.

Hence, the consequence that it is true that **S** or it is false that **S** follows immediately by *Modus Ponens* from (3) and (4), a consequence that contradicts the assumption that **S** is truth-valueless. Since **S** is arbitrary, the conclusion holds generally for any instance of **A** in (1).

To sum up, the notion of validity as merely truth-preserving inclines one toward a conception of logical truth that presumes nothing about

the truth-values of statements. On the other hand, the classical codification of logical truths plus the Tarski adequacy condition demand
acceptance of an account of logical truth that definitely excludes truth-
valueless statements.

The conflict just described probably surprises few if any. But reaction to it runs the gamut from the concern it evokes in those who
believe that logic is merely a tool and should not prejudge philosophical questions to the ho-hum response of those who have adjusted to
the facts of logical life.

One way out of the present state of affairs would be to reject, or to
modify, Tarski's Convention T, the adequacy condition for truth.[5] In
the 1960s this course appeared to have had few if any takers. Probably
this was because many most logicians at the time were inclined to believe that an adequate semantics for classical statement logic allowing
for truth-valueless statements simply was not possible.

It is not hard to find cases. For example, one finds the attitude
expressed by Wesley Salmon in his essay, 'Verifiability and logic'.[6] In an
argument designed to show the question-begging character of Ayer's
account of cognitive meaningfulness, Salmon writes:

> Very briefly, the situation is this. Cognitive statements are taken to be just
> those statements that are either true or false. Statements that are either true
> or false are the admissible substituends for the variables of truth-functional
> logic. But, for those statements that are neither analytic nor self-contradictory,
> a statement is cognitive if and only if it is empirically verifiable. The test for
> empirical verifiability involves using a statement as a substituend for a truth-
> functional variable (or larger expression involving variables) in the premise
> of a deductive argument. *This procedure is logically permissible only if the statement
> in question is either true or false, which is precisely the question at issue.*[7]

However, van Fraassen's semantical investigations in the 1960s provided a way of characterizing logical truth that did not do violence
to the conception of validity as merely truth preserving; that is, truth-
valueless statements were not excluded but the classical codification

---

[5] Bas C. van Fraassen, 'Singular Terms, truth-value gaps, and free logic,' *Journal of Philosophy*, 63 (1966), pp. 481–495.

[6] W. Salmon, 'Verifiability and Logic', in *Mind, Matter, and Method: Essays in Philosophy and Science in Honor of Herbert Feigl* (eds. P. Feyerabend and G. Maxwell), University of Minnesota Press, Minneapolis, 1967, p. 359.

[7] My italics. Salmon himself kindly brought this passage to my attention.

of statement logic was nevertheless retained. The implication of this proposal is that the classical logic of statements is neutral with respect to whether there are or are not truth-valueless statements. To be sure, some ideas must be modified, the major one being a modification of Tarski's adequacy condition. That this is not as drastic as might appear at first glance will emerge shortly.

2. A REVISED ACCOUNT OF LOGICAL TRUTH

In the classical logic of statements, logical truth is explicated with the help of the notion of an assignment. Typically an *assignment* is a total function from the simple (or atomic) to the set of truth-values $\{\mathbf{T}(\text{ruth}), \mathbf{F}(\text{alsity})\}$. A *classical valuation* ($\mathbf{Val_c}$) is a total function from the entire set of statements of the language to the same set of truth-values that agrees with the truth-values assigned to atomic statements; where α is an assignment, $\alpha(\mathbf{A}) = \mathbf{Val_c}(\mathbf{A})$ if $\mathbf{A}$ is atomic. $\mathbf{Val_c}$ yields a definition of truth (falsity) for the logic of statements.

Van Fraassen introduced the notion of a *supervaluation* ($\mathbf{Val_s}$) as a partial function from classical valuations to the set of truth-values $\{\mathbf{T}, \mathbf{F}\}$. It may defined as follows.

> $\mathbf{D_s}$ For all statements $\mathbf{A}$ of the language $\mathbf{L}$, $\mathbf{Val_s}(\mathbf{A}) = \mathbf{T}$ if and and only if for all $\mathbf{Val_c}$, $\mathbf{Val_c}(\mathbf{A}) = \mathbf{T}$, $\mathbf{Val_s}(\mathbf{A}) = \mathbf{F}$ if and only if for all $\mathbf{Val_c}$, $\mathbf{Val_c}(\mathbf{A}) = \mathbf{F}$, and otherwise $\mathbf{Val_s}(\mathbf{A})$ is undefined.

Letting the *admissible* valuations be supervaluations, logical truth is then defined as truth in all supervaluations. The definition $\mathbf{D_s}$ assures that the set of logical truths defined via classical valuations is exactly the same as those defined in terms of supervaluations, but without the presumption that the component statements so defined must have truth-values.[8] This is the case because a supervaluation picks out what is common in all classical valuations, but assigns nothing in those cases where there is truth-value variation among classical valuations. (See Chapter 8, section 3.) The intuitive idea is that if a statement is regarded as truth-valueless, then if it were to be assigned a truth-value in a classical valuation, the assignment would be arbitrary; it could, with equal justification, be assigned $\mathbf{T}$ or be assigned $\mathbf{F}$.

---

[8] See Ermanno Bencivenga, Karel Lambert and Bas van Fraassen, *Logic, Bivalence and Denotation* (second edition), Ridgeview, Atascadero, California (1991), pp. 43–47.

Before turning to the implications of these ideas for the logic of microphysics, two items have to be discussed briefly; the first has to do with the philosophical neutrality of the classical codification of statement logic, and the second with Tarski's adequacy criterion.

Van Fraassen's definition of logical truth via supervaluations is neutral on the question of whether or not the components of a logical truth, atomic or compound, have a truth-value; thus (1) will count as logically true even if **A** is truth-valueless. On the other hand, (1) clearly is logically true when **A** has a truth-value. Accordingly, the present explication of logical truth is neutral with respect to the question of whether or not there are truth-valueless statements. Since it is provable that a statement is logically true *qua* supervaluations if and only if it is a theorem of classical statement logic,[9] it follows that the classical codification of logic is neutral with respect to the truth-value status of statements.

There is a proviso, however; Tarski's adequacy condition for truth must be modified. For recall that it was by means of it that any statement of the form (1) was shown to imply that its component statements are true or false. So a notion of logical truth that conforms to the idea of validity as merely truth-preserving requires a modification of Tarski's adequacy condition, *if* one wishes to retain the classical codification of statement logic. This implication should not alarm; it would be more accurate to view the present analysis as revealing a presumption latent in Tarski's adequacy condition rather than as requiring a drastic alteration in the notion of truth. For all this analysis demands is that, however one discriminates between the truth-valueless and the truth-valued, the Tarski adequacy criterion holds for any statement meeting the standard for being truth-valued. For example, in van Fraassen's own semantics for free logic, every atomic non-identity statement containing a singular term without existential import is taken to be truth-valueless.[10] So, in these cases, the adequacy criterion might be modified as follows: If **A** is an atomic non-identity statement containing a singular term **t** then (**X** is true if and only if **A**), where **X** is the name of

[9] See Bas van Fraassen, 'The completeness of free logic', *Zeitschrift für mathematische Logik und Grundlagen der Mathematik*, 12 (1966), pp. 219–224. In this essay supervaluations are called 'state-models'.

[10] Op. cit., Bas van Fraassen, *The Journal of Philosophy* (1966).

A... etc ... etc, *provided* $\exists x(x = t)$, where $\exists x(y = t)$ is the object language correlate of **t** has existential import.

3. TRUTH-VALUE GAPS VERSUS ANOTHER TRUTH VALUE

So far certain statements have been said to be truth-valueless rather than neither true nor false. The reason is that too often the expression 'is neither true nor false' suggests a third truth-value – some called it the value Middle, others the value Meaningless or Absurd. The difference, semantically speaking, is that between assigning something to a statement and not assigning anything; one might think of the expression 'no truth-value assignment is made' as the formal counterpart of Strawson's informal aphorism "the question concerning the truth-value of the statement does not arise."

The distinction is not idle because truth-value gap semantics for the logic of statements, as developed by van Fraassen, in contrast to conventional classical three valued semantics, is *non-extensional*; co-valent statements do not substitute everywhere *salva veritate*. To see this, note that all instances of (1) are logically true (hence true) in supervaluational semantics even if the instance of **A** has no truth-value. Label such an instance

S.

Let

T

be co-valent with **S**, hence truth-valueless. Then though

S or it is not the case that **S**

would be true,

T or it is not the case that **S**

would not because in some classical valuations it would be true (those in which **T** is true) and in some false (those in which both **S** and **T** are false).[11]

---

[11] In a recent paper, Thomasz Bigaj, 'Three-valued logic, indeterminacy and quantum mechanics', *Journal of Philosophical Logic*, 30 (2001), pp. 97–119, has resurrected supervaluations as a foundation for quantum mechanics. Despite its ingenuity, it overlooks

The more common interpretation of the expression 'is neither true nor false' as representing a third truth-value rather usually is associated with a non-classical codification of statement logic. Thus, Goddard, in his work on formal theories of meaningfulness, treats Ryle's assertion that category mistakes are neither true nor false as requiring a third truth value and thus a non-classical codification of significance.[12]

More to the point, the issue of how to interpret the expression 'is neither true nor false' has implications for the problem of the logic of microphysics. In 1957, Hilary Putnam[13] argued that the expression 'is neither true nor false', construed as representing a third truth-value, Middle, could be applied gainfully to certain physical statements of the form 'Object **b** has position **G** at time **t**'. He believed a three-valued logic had at least one profound effect in microphysics: It allowed one, as Reichenbach had previously noted, to hold consistently both to the laws of quantum physics and to the principle that there is no velocity greater than the speed of light. But, he asserted, in classical two-valued logic, the principle that there is no velocity greater than the speed of light is incompatible with the laws of quantum mechanics, given the meaningfulness of all statements of the form 'Object **b** has position **G** at time **t**'.

Isaac Levi[14] and Paul Feyerabend[15] objected to Putnam's view recommending that the middleman be eliminated in the logic of microphysics. Levi complained that there is no real choice between a two-valued logic that recognizes unknown truth-values and the three-valued analogue that assigns to exactly the same statements a third truth-value, namely, Middle.[16]

Feyerabend did not challenge Putnam's belief that the truth-value Middle can be given a coherent interpretation, but he rejects Putnam's

---

the current essay (originally published in 1969) and van Frassen's 'The labyrinth of quantum logics' (1972) which summarises the results of the current essay (see below, footnote 22), and obfuscates the essential difference between truth-value gaps and three-valued logic.

[12] Leonard Goddard, 'Predicates, relations and categories,' *Australasian Journal of Philosophy*, *44* (1966), pp. 139–171.

[13] Hilary Putnam, 'Three-valued logic', *Philosophical Studies*, 5 (1957), pp. 73–80.

[14] Issac Levi, 'Putnam's three truth values', *Philosophical Studies*, 5 (1959), pp. 65–69.

[15] P. Feyerabend, 'Reichenbach's interpretation of quantum-mechanics', *Philosophical Studies*, 4 (1958), pp. 49–59.

[16] Op. cit., 'Putnam's three truth values' (1959), p. 68.

three-valued proposal as being but the most modern of "the sly procedures which have been invented for the purpose of saving an incorrect theory in the face of refuting evidence." For the statement that there is no velocity greater than the velocity of light "is a well-corroborated statement of physics."[17] Moreover, he finds Putnam's allegation of an incompatibility in classical two-valued logic between the laws of quantum mechanics and the principle that no velocity exceeds that of the speed of light specious.[18] However, says Feyerabend, Putnam's three-valued proposal does support another incompatibility: The laws of quantum mechanics are incompatible with the principle that every entity possesses always one property out of each of the set of mutually exclusive classical categories. This principle implies that every entity has both a well-defined position and a well-defined momentum. Acceptance of Putnam's proposal, he asserts, would require us to acknowledge the propriety of the *false* principle that every entity always has one out of every pair of mutually exclusive classical properties.[19]

Feyerabend also takes Putnam and Reichenbach to task for believing that the three-valued way out of the difficulties occasioned by the physics of the minute is superior to Bohr's proposal to treat the offensive statements as "meaningless."[20] He accuses Putnam and Reichenbach of *apriorism* in their belief that the statements Bohr calls "meaningless" indeed "have a very clear cognitive use." He further argues that the Putnam-Reichenbach proposal fails to satisfy the Reichenbach standard of adequacy that "every law of quantum mechanics should have either the truth-value True or the truth-value False, but never the truth-value "Indeterminate." Specifically, Feyerabend argues that, given the Putnam-Reichenbach three-valued proposal, "every quantum-mechanical statement containing noncommuting operators can only possess the truth-value [Middle]."[21]

Feyerabend's objection that quantum-mechanical statements containing noncommuting operators get the truth-value Middle in the Putnam-Reichenbach approach, thus violating a Reichenbachian

[17] Op. cit., 'Reichenbach's interpretation of quantum-mechanics', (1958), p. 50.
[18] Ibid., p. 53.
[19] Ibid., p. 51.
[20] Ibid., p. 53.
[21] Ibid., p. 54.

standard of adequacy, however, is *not* a legitimate complaint against the position that certain elementary quantum mechanical statements are truth-valueless as opposed to Middle.[22]

4. RECONSRUCTING REICHENBACH'S INTERPRETATION OF ELEMENTARY MICROPHYSICAL STATEMENTS

Consider the following pair of statements,

(5)   There are mountains on the other side of the moon,

and

(6)   The Queen of the United States wants to marry a man named 'Harry'.

Before rocketships, (5) was a standard example of a statement that had a truth-value, but whose particular truth-value was unknown. Still its truth (or falsity) was potentially verifiable (or falsifiable). However, (6) is very different. Because it is neither potentially verifiable nor potentially falsifiable, it would be misleading to say that its truth-value is unknown. Putnam's remarks about the truth-value Middle seem to point to some such distinction, for he says that it does not mean "unknown truth-value." Further, he suggests that, under certain conditions, one can know by virtue of a physical law and certain observational data that a statement like 'Object **b** has position **G** at time **t**' can never be falsified or verified.[23] This would seem to put statements about an object's position in the same group as statements about the Queen of the United States. This of course does not mean that there is no distinction at all between statements about the Queen of the United States and object **b**; the good Queen does not exist, but presumably object **b** does. If 'neither true nor false' means 'truth-valueless', then it does not follow that treating elementary microphysical statements as neither true nor false presumes a non-classical codification of statement logic. For those, like Putnam and Reichenbach, who find sense in such statements, this approach to elementary microphysical statements would surely be

[22] See Bas van Fraassen, 'The labyrinth of quantum logics', *Boston Studies in the Philosophy of Science*, XIII (1972), pp. 224–254.
[23] Op. cit., 'Three-valued Logic' (1957), p. 75.

more palatable than the Bohr approach. His approach, in effect, excludes such statements as legitimate instances of classical two-valued schemata.

To amplify, consider van Fraassen's application of supervaluations to free logic.[24] The conditions there under which an atomic statement is assigned true, false, or nothing in the supervaluation **Val$_s$** may be summed up as follows:

Let **F(t)** represent an arbitrary atomic nonidentity statement containing 't' as singular term. Then

    (i) **F(t)** is true if **t exists** and is a member of the class of things that are **F**.

   (ii) **F(t)** is false if **t** exists but is not a member of the class of things that are **F**.

  (iii) **F(t)** is truth-valueless if **t** does not exist.

For when **t** exists and is a member of the class of things that are **F**, **F(t)** is true in all classical valuations, hence true in the supervaluation **Val$_s$**. Similarly, when **t** exists but is not a member of the class of things that are **F**, **F(t)** is false in all classical valuations, hence false in the supervaluation **Val$_s$**. Finally, when **t** does not exist, **F(t)** is false in some classical valuations and true in others, hence truth-valueless in the supervaluation **Val$_s$**.

In analogous fashion truth-conditions can be laid down for elementary microphysical statements such that one can tell which statements will be assigned true, false, or nothing in a given supervaluation.[25] Thus let **G(b,t)** be short for '**b** has position **G** at time **t**'. Then in **Val$_s$**,

    (i) **G(b,t)** is true if **b** has position **G** at time **t**.

   (ii) **G(b,t)** is false if **b** has some position other than **G** at time t.

  (iii) **G(b,t)** is truth-valueless if **b** has a property (for example momentum **M**) which is quantum theoretically incompatible with having a definite position at time **t**.

---

[24] Op. cit., 'Singular terms, truth value gaps and free logic' (1966).

[25] Op. cit., 'The labyrinth of quantum logics' (1968). For a more complete account of the supervaluational treatment of elementary microphysical statements, see pp. 235–237.

Care must be taken to distinguish between

(7) It is always the case that **G(b,t)** **V** ~**G(b,t)** is true,

and

(8) It is always the case that **G(b,t)** is true or that ~**G(b,t)** is true.[26]

From the previous discussion it ought to be plain that, under the present interpretation of elementary microphysical statements, (7) does **not** imply (8). And in fact (8) is not always the case, for example, when **b** has a definite momentum at time **t**; that is, under such circumstances, neither **G(b,t)** nor ~**G(b,t)** is true. This is captured in (iii) above for **G(b,t)** in the supervaluation **Val$_s$**.

Consider now the classical principle **C**, the principle that "each entity possesses *always* one property out of each [classical] category." **C** implies (7). Reichenbach apparently thought that it also implied (8), because he thought that (7) implies (8).[27] Thus he was led to what surgeons call the "heroic treatment" of rejecting the classical codification of statement logic. Given Reichenbach's ultimate rejection of both (7) and (8), it is, indeed, hard to see how Reichenbach could retain the classical principle **C**. The most that is claimed for the present interpretation of elementary microphysical statements is that it represents a reasonable reconstruction of Reichenbach's position, given his desire to maintain **C**.

The present reconstruction of the Reichenbach-Putnam position on elementary microphysical statements should be congenial to those who find Bohr's approach unacceptable – at least as interpreted by Reichenbach. Besides having a definite if-it-offends-thee-pluck-it-out aura about it, the Bohr approach is in some respects more extreme than the Reichenbach-Putnam position. Bohr's position is that a statement such as '**b** has position **G** at time **t**' is "meaningful" only if certain physical conditions are realized. The upshot is that elementary microphysical statements are allowed to be substituends for the schematic letters in the schemata of classical

[26] (8), of course, is an instance of the principle of *Bivalence*, a principle that fails in supervaluational semantics.

[27] Op. cit., 'Reichenbach's interpretation of quantum-mechanics' (1957), p. 51.

statement logic only if certain measurement conditions are realized. There is a striking similarity here to the attitude of those who deny the conditional

(9)   If everything exists, so does 1/0

the status of a counterexample to the validity of the logical law called *Specification* on the ground "that '1/0 does not exist' is "ill formed" or "meaningless." But surely this is the counsel of the ostrich. It amounts to burying one's head in the sand in the face of a threat to a long-standing theory. It suggests that no new theories would ever be necessary because there would really be no conflicting instances to the old theory to explain; the purported counter-examples can simply be disqualified as nonsensical.

Feyerabend's accusation of *apriorism* to Putnam's assertion that such sentences as 'Object **b** has position **G** at time **t**' have a clear cognitive use is unconvincing. His assertion that "in our search for better theories we frequently discover that situations we thought would obtain universally do in fact exist only under special conditions, which implies that the properties of these situations are applicable in those conditions only"[28] is weak. Often it is the case that situations we think to obtain universally don't – as in the "discovery" that $\exists x(x = t)$ holds only when **t** is a term with existential import. But that doesn't imply that it is only applicable in situations in which **t** does have existential import. The resulting statements *do* have a clear cognitive use – as in the case of $\exists x(x = \text{Vulcan})$ – and are false.

Turning to assessment of Levi's objection to Putnam's third truth-value Middle that there is no distinctive answer to the question: What is this thing called Middle?. Levi's argument loses its force, by default if you will, there being no "thing" which a statement is assigned when it is said to be truth-valueless. Accordingly the present reconstruction of the Putnam-Reichenbach position is not vitiated by Levi's arguments against three-valued logic.

The most important philosophical implication of the present reconstruction of Reichenbach's interpretation of elementary microphysical statements is this: Which truth-value, if any, is to be assigned to certain elementary microphysical statements is separable from the question

---

[28] Ibid., p. 56.

of whether microphysics requires a non-classical logic. For even if no truth-value is assigned to certain microphysical statements it is still possible to define 'valid', 'logical truth', and so on, in such a way that all classically valid arguments in the logic of statements remain valid, and all classical logical truths therein remain logically true. This implication runs against the grain of the historical tradition that teaches that classical logic dictates a particular interpretation of microphysical statements, and insofar as the required interpretation runs afoul of the laws governing the physics of the minute, classical logic is wanting. The force of the present reconstruction of Reichenbach's position has been rather that classical logic is neutral with respect to the interpretation of elementary microphysical statements. It is no more incompatible with rejection of the principle that every entity always has a determinate position than it is with acceptance of that principle. At the semantic level, this shows up in the fact that the truth of (7) is independent of the truth of (8).

5. DODGING SOME BULLETS

Some hold that it is non-intuitive that every statement of the form $F(t)$ V $\sim F(t)$, where F is a predicate and t is a singular term without existential import, is true let alone logically true. In a personal conversation Michael Scriven once suggested to me that 'The greatest natural number is prime or it is not' is such a case. So, in this respect at least, the present reconstruction of Reichenbach's position is inferior to Reichenbach's original approach via three-valued logic. In Reichenbach's original approach Scriven's example would not be logically true and could indeed be assigned the value Middle consistent with "ordinary intuition".

Three replies come to mind. First, intuition is a questionable guide in this matter. Recall that the characterization of logical truth given earlier was motivated by a certain basic intuitive consideration explained in the first two sections section of this essay. And *that* intuitive guide does yield the result that a statement like 'The greatest natural number is prime or it is not' is logically true.

Secondly, it is not necessarily the case that adoption of a logic in which Excluded Middle fails for atomic statements containing singular terms without existential import requires adoption of such a logic in microphysics. For, in the latter case, presumably **b** in 'Object **b** has **G** at time **t**' exists. Hence the reasons for calling it truth-valueless will be quite different.

Thirdly, even granting the intuitive non-truth of (1) when its components are neither true nor false, who is to say that subsequent developments in microphysics and in its logic might not dictate an alteration in our intuitions more in line with the present interpretation of elementary microphysical statements? The success of Newtonian physics, after all, required an alteration in ordinary (Aristotelian) intuitions about the behavior of physical objects, an analogy brought to my attention by Stephen Korner.

Milton Fisk, and independently Wesley Salmon, have suggested informally that logical laws can be regarded as truth-valueless when they have truth-valueless components without thereby having to abandon the classical codification of statement logic. The trick is to interpret 'logically true' as meaning simply 'is never assigned false'. Then, for example, any statement of the form in (1) will be logically true, though there will be instances of this schema that are truth-valueless. Fisk suggests that this latter approach would be more acceptable to those who do not find the subjunctive "Ignoring existence presuppositions, if the components of instances of (1) *were* assigned truth-values, those instances *would* be true" an adequate basis for assigning truth to an instance of (1) when its components are in fact truth-valueless.

This complaint evokes two responses. First, Fisk's suggestion has the following unpalatable consequence. Assuming logical truth as a measure of validity, Fisk's alteration in the concept of logical truth requires a similar alteration in the concept of validity; validity must now be construed as 'non-falsehood preserving'. Fisk's suggestion has the effect of letting certain conditionals be logical truths which though not always assigned **T** nevertheless are never assigned **F**. But then there may be valid arguments with, say, true premises and a truth-valueless conclusion. This destroys the usefulness of inference as an instrument of proof in the customary sense of establishing the truth of some claim.

Secondly, it is not true that the Fisk interpretation of logical truth necessarily yields the classical codification of statement logic. The rule of *Modus Ponens*, a crucial rule in classical statement logic, is truth preserving, but it is not non-falsehood preserving in the present semantical development. Thus, let **S** be truth-valueless and let **S*** be assigned **F**. Then **S** is non-false and **S** ⊃ **S*** is non-false, though **S*** is not non-false.

Finally, it is readily acknowledged that the present reconstruction of the Reichenbach interpretation of microphysics, and, indeed, Reichenbach's original interpretation, provides at best an analysis of the surface structure of the set of elementary microphysical statements. Supplementation will be necessary to get at the deeper structure of this set. But the major purpose here has been foundational in a more philosophical sense, and in this respect it offers some evidence for the belief that quantum mechanics does not require special logics called 'quantum logics'.